Theory and Practice of STEAM Education in Japan

With unique insights into the potential power of Japan's STEM education, Isozaki and his team of contributors share multiple perspectives on STEM education theory and practices in Japan.

Examining how Japan has become an economic superpower based on scientific and technological innovations, this book provides a particular focus on the theoretical and practical analysis of STEM education from historical and comparative perspectives. Additionally, it links the theory and practice of STEM education from primary education to teacher education at universities across Japan and considers both societal and individual needs in advancing STEM literacy. Chapters are written by researchers from a diverse range of fields in education, including science, mathematics, technology, and pedagogy. The book also offers practical teaching tools and materials for teacher education and assessment to promote STEM literacy in students so that they are able to address local and global socio-scientific issues in a real-world context.

Covering a wide spectrum of STEM education, this book provides valuable insights and practical suggestions, from a Japanese perspective, for academic researchers, policymakers, and educators who are interested in STEM education.

Tetsuo Isozaki is a professor of Graduate School of Humanities and Social Sciences at Hiroshima University, Japan.

Routledge Research in Education

This series aims to present the latest research from right across the field of education. It is not confined to any particular area or school of thought and seeks to provide coverage of a broad range of topics, theories and issues from around the world.

Developing a Didactic Framework Across and Beyond School Subjects
Cross- and Transcurricular Teaching
Edited by Nina Mård and Søren Harnow Klausen

Reimagining Boredom in Classrooms through Digital Game Spaces
Sociomaterial Perspectives
Noreen Dunnett

On the Theory of Content Transformation in Education
The 3A Methodology for Analysing and Improving Teaching and Learning
Tomáš Janík, Jan Slavík, Petr Najvar and Tereza Češková

Living Educational Theory Research as an Epistemology for Practice
The Role of Values in Practitioners' Professional Development
Jack Whitehead and Marie Huxtable

Empowering Teachers for Equitable and Sustainable Education
Action Research, Teacher Agency, and Online Community
Maria Teresa Tatto

Theory and Practice of STEAM Education in Japan
Edited by Tetsuo Isozaki

For more information about this series, please visit: www.routledge.com/Routledge-Research-in-Education/book-series/SE0393

Theory and Practice of STEAM Education in Japan

Edited by Tetsuo Isozaki

Routledge
Taylor & Francis Group

LONDON AND NEW YORK

First published 2024
by Routledge
4 Park Square, Milton Park, Abingdon, Oxon OX14 4RN

and by Routledge
605 Third Avenue, New York, NY 10158

Routledge is an imprint of the Taylor & Francis Group, an informa business

British Library Cataloguing-in-Publication Data
A catalogue record for this book is available from the British Library

ISBN: 978-1-032-49190-5 (hbk)
ISBN: 978-1-032-49191-2 (pbk)
ISBN: 978-1-003-39254-5 (ebk)

DOI: 10.4324/9781003392545

Typeset in ITC Galliard Pro
by KnowledgeWorks Global Ltd.

Contents

Contributors

Kiichi Amimoto is a professor at Seinan Gakuin University, Japan.

Yusuke Endo is an assistant professor at The University of Tsukuba, Japan.

Terumasa Ishii is an associate professor at Kyoto University, Japan.

Takako Isozaki is a professor at The University of Toyama, Japan.

Tetsuo Isozaki is a professor at Hiroshima University, Japan.

Takashi Kawakami is an associate professor at Utsunomiya University, Japan.

Keiichi Nishimura is a professor at Tokyo Gakugei University, Japan.

Susumu Nozoe is a professor at The University of Miyazaki, Japan.

Tadashi Ohtani is a professor at Tokyo Gakugei University, Japan.

Ryugo Oshima is an assistant professor at Chiba University, Japan.

Moegi Saito is a lecturer at The University of the Sacred Heart, Tokyo, Japan.

Ko Tomikawa is a professor at Hiroshima University, Japan.

Introduction

Tetsuo Isozaki

Many articles on science, technology, engineering, and mathematics (STEM) education have been published worldwide since the beginning of the millennium. Marginson (2015) categorized nations into three types based on cultural region: English-speaking nations, Western Europe, and the Post-Confucian countries. Among these categories, Japan is a Post-Confucian country. As Marginson (2015) argued, compared to students from English-speaking nations and Western Europe, those in Asian countries, especially "the four Asian tigers" (South Korea, Taiwan, Hong Kong, and Singapore) and mainland China and Japan, often obtain higher scores in international tests such as the Trends in International Mathematics and Science Study and Programme for International Student Assessment. While Asian countries have long been influenced by the West, unlike China and South Korea, Japan has traditionally adopted different approaches to education since its modernization in the mid-19th century. This implies that education is not simply imported from the West but is recontextualized based on the country's social and cultural context. For example, Isozaki (2014) noted that science education, called *rika* in Japanese (e.g., Ogawa, 2015), was recontextualized in Japan's socio-cultural context in the late 19th century. Consequently, *rika* has characteristics that differ from those of Western science education. In addition, mathematics education in elementary schools, called *sansu* in Japanese, can be considered an integration of both Japanese mathematics, called *wasan*, and Western mathematics.

Although many educators, researchers, and politicians are interested in STEM or STEAM (STEM + Arts or Liberal Arts) education, the theory and practice thereof in schools in Japan are still being explored. As such, there are currently no uniform definitions of STEM and STEAM. In fact, the government of Japan has acknowledged diverse international definitions of STEAM. However, STEAM is officially viewed as cross-curricular education through which students use the learning gained in each subject to discover and solve real-world problems (e.g., Ministry of Education, Culture, Sports, Science and Technology [MEXT] (Central Council for Education), 2021). If education is considered a social and cultural product, it is likely influenced by a nation's culture, government policies, and educational traditions.

DOI: 10.4324/9781003392545-1

There are two STEM education policies worldwide. First, if the purpose of STEM education is the development of human resources to supply future scientists and engineers to the STEM fields, then it would likely have emerged in the West and Japan in the 19th century when modern schooling was becoming more prolific, long before the term "STEM" was coined. Second, if the purpose of STEM education is to ensure that all future citizens are sufficiently STEM-literate to solve the socio-scientific issues arising in knowledge-based societies, then it would have originated during the transition from the third to fourth industrial revolutions. For example, the Imperial Universities established by Japan's government in the second half of the 19th century and first half of the 20th century, which were more weighted toward faculties of science, medicine, and engineering than non-scientific faculties, aimed to develop human resources for leadership positions in STEM fields. While the recent policies of Japan's government and business communities regarding the STEAM disciplines have enhanced the development of human resources, they also advocate the importance of employing STEAM education to educate citizens (e.g., Japan Business Federation (*Keidanren*), 2014; (MEXT [Central Council for Education], 2021). At the start of the millennium, MEXT and other ministries—including the Ministry of Economy, Trade and Industry; the Ministry of Agriculture, Forestry and Fisheries (MAFF); and the Japan Science and Technology Agency—provided various programs related to STEM and STEAM (hereafter, "STEAM" is abbreviated as "STEM" unless otherwise specified) education in both formal and informal education settings. Essentially, some believe STEM targets excellence (STEM for excellence), while others believe that STEM education should ensure STEM literacy for all citizens (STEM for all). Consequently, we now find ourselves in "the age of STEM education."

In this context, limited research in English demonstrates the potential of STEM-related fields for education researchers in Japan. For example, Teo et al. (2022) investigated STEM education in Asian countries; however, they did not include Japan in their study. To address this gap, this book provides an opportunity for mid-career and young researchers to share the unique features of Japan's STEM education with a global audience. As such, this book provides an important opportunity for Japanese education researchers to disseminate information and demonstrate the potential power of Japan's STEM education. This is just one reason we are keen to highlight the Japanese context in this book.

Within STEM education, as STEM-literate future citizens, students need to be able to combine, integrate, and interlink diverse knowledge and skills to create solutions to local and global socio-scientific issues in a real-world context. To this end, they must learn how to appropriately apply what they have studied in each STEM subject, as well as other subjects and activities, to a general STEM scenario. The contributors to this book come from a wide range of education fields, including science, mathematics, technology, home economics, and pedagogy. Consequently, this book covers the entire spectrum of STEM education, including theory and suggestions for practice.

In Chapter 1, Tetsuo Isozaki analyzes the history of each STEM subject to identify its nature, with reference to economic dialogues from the perspective of the nation and economic growth. The author highlights that to advance STEM literacy in Japan's educational context, it is important to consider both societal needs and individual perspectives. Furthermore, the chapter argues that STEM for excellence should be positioned as part of STEM education for all.

In Chapter 2, Tadashi Ohtani examines the roles of science, technology, engineering, arts or liberal arts, and mathematics as components of STEAM according to the framework of subjects in Japanese education policy. With reference to the United States, the author describes the relationship between the "A" and "E" in STEAM according to the framework of subjects in Japan.

Traditionally, Japanese education researchers and policymakers have monitored Western education trends, especially in the United States, Germany, and the United Kingdom, and recontextualized them to fit Japan's context. Currently, the trends of STEM education in the United States are being introduced and compared with those in Germany and the United Kingdom. Therefore, in this book, Yusuke Endo and Susumu Nozoe each adopt a comparative education approach to focus on Germany in Chapter 3 and the United Kingdom in Chapter 4, respectively. Both authors received their Ph.D. degrees in science education with comparative education. Through comparisons with the West, these chapters highlight the characteristics of Japan's STEM education and aspects that can be learned from Germany and the United Kingdom.

In Chapter 5, Terumasa Ishii explores an assessment method for fostering STEM literacy in Japan, based on performance assessment theory in Japan. STEM education in the United States has been implemented as cross-curricular learning in which knowledge and skills are used in complex ways in a real-world context. Accordingly, the author summarizes the development of theories and practices that use performance assessment in science, mathematics, technology, and integrated learning time clarifying the nature of assessment that fosters STEM literacy.

Effective collaborative learning in STEM education should be designed based on hypotheses of the learning process derived from evidence of student learning in the classroom. In Chapter 6, Moegi Saito identifies measures for assessing constructive interactions that foster a deeper understanding of each student. Conducting empirical research with her colleagues, she considers how these measures can be used to ensure teachers fully understand collaborative learning and verify their instructional design hypotheses based on students' learning processes.

In Japan, MEXT and MAFF provide STEAM education. In Chapter 7, Takako Isozaki and Tetsuo Isozaki discuss the history of food and nutrition education (*shokuiku*), incorporating the concept of food literacy. They propose addressing the possibility of multidisciplinary and interdisciplinary approaches for teaching *shokuiku*, in which home economics teachers, supported by nutrition teachers, play a central role that needs to be collaboratively addressed by schools and teachers.

Educating pre-service teachers to impart lessons incorporating a STEM education approach that emphasizes the relevance of each STEM subject and domain in every lesson is a major challenge in Japan as well as other countries. In Chapter 8, Takashi Kawakami and Keiichi Nishimura present a Japanese example of pre-service teacher education for STEM activities, which combines complementary data modeling and design processes. They also discuss the process and content integration of the pre-service teachers' activities and implications for STEM pre-service teacher education.

In STEM education in Japan, it is necessary to develop integrated learning activities at the upper secondary school level. However, it is also critical to examine and enhance learning in each subject up to lower secondary school to evaluate whether it leads to integrated learning at the upper secondary school level. In Chapter 9, Ryugo Oshima discusses strategies for enhancing each STEM subject in lower secondary school from a science perspective to enrich STEM education at the upper secondary school level.

Finally, in Chapters 10 and 11, Kiichi Amimoto and Ko Tomikawa, both scientists, report on engaging in teacher education, especially in the development of teaching materials for use in STEM education. In Chapter 10, Amimoto discusses the features of the discipline and learning of chemistry and the methodology for designing teaching materials for Japanese-style inquiry activities. He presents examples of cross-disciplinary inquiry activity based on STEM concepts. In Chapter 11, Tomikawa overviews research and educational practices related to STEM education focused on biological sciences in Japan. He offers examples of STEAM teaching materials using biotechnology and GIS. They also review the application of the STEM approach in learning chemistry and biology, respectively.

Acknowledgments

This book has been conducted as a research result supported by the Japan Society for the Promotion of Science KAKENHI Grant Numbers JP20K20832, JP21H00919 (JP23K20744).

References

Isozaki, T. (2014). The organisation and the recontextualization of *rika* (school science) education in the second half of the nineteenth century in Japan. *Science & Education, 23*, 1153–1168. https://doi.10.1007/s11191-013-9615-4

Japan Business Federation (*Keidanren*). (2014). *Rikōkei jinzaiikusei senryaku no sakutei ni mukete (Towards the formulation of science and technology human resource development strategies). Keidanren* [in Japanese]. Retrieved from https://www.keidanren.or.jp/policy/2014/013.html

Marginson, S. (2015). What international comparison can tell us. In B. Freeman, S. Marginson, & R. Tyler (Eds.), *The age of STEM: Educational policy and practice across the world in science, technology, engineering and mathematics* (pp. 22–32). Routledge. https://doi.org/10.4324/9781315767512

Ministry of Education, Culture, Sports, Science and Technology (MEXT) (Central Council for Education). (2021). *"Reiwa no nihongata-gakkōkyōiku" no kouchiku wo mezashite: Subete no kodomotachi no kanousei wo hikidasu, kobetsusaitekinamnabi to kyōdoutekina manabi no jitugen (tōshin) (Toward the construction of "Japanese style school education": Optimal individualized learning and collaborative learning that bring out the potential all children (report)).* MEXT [in Japanese]. Retrieved from https://www.mext.go.jp/content/20210126-mxt_syoto02-000012321_2-4.pdf

Ogawa, M. (2015). Rika. In R. Gunstone (Ed.), *Encyclopedia of science education* (p. 840). Springer.

Teo, T. W., Tan, A-L., & Teng, P. (Eds.). (2022). *STEM education from Asia: Trends and perspectives.* Routledge. https://doi.org/10.4324/9781003099888

1 What can we learn from the history of STEM subjects in Japan?

Tetsuo Isozaki

Introduction

Background

This chapter explores the historical nature of each STEM subject—science, technology, engineering, and mathematics—in Japan, aiming to develop STEM literacy for all future citizens. The chapter reviews the period from the second half of the 19th century until today. Japan has learned from the West since the 19th century; however, it did not merely imitate Western practices but instead recontextualized its learnings for the Japanese cultural and societal contexts.

From the 1990s, STEM education became an urgent political agenda in both the West and the East. Nevertheless, as Breiner et al. (2012) argued, there is no agreement on the definition of STEM. Some government documents, such as Sir Gareth Roberts' Review (Roberts, 2002) in the United Kingdom (U.K.), reports by the United States (U.S.) Department of Commerce, Economics, and Statistics Administration (2011), and by the Committee on STEM Education of the National Science and Technology Council (2018), have stated that STEM education is linked to competition, helping maintain a nation's global position and economic growth through the power of science and technology.

Education issues, particularly those related to science, mathematics, technology education, have affected economic discourse, especially post-World War II (WWII). Japan has regarded education as an essential vehicle to build the nation/state and achieve economic growth since the mid-19th century. This link between human resource development, primarily in the fields of science and technology, and economic growth was behind the assertion of the importance of STEM education. Nevertheless, as Bybee (2010) highlighted, we must not forget that another crucial aim of STEM education is to develop STEM literacy for all.

Additionally, in recent years, the term "STEM" has been extended to "STEAM." The meaning and role of the "A" within STEAM is complicated and could represent art, arts, or liberal arts. In one extreme view, the "A" could, in a sense, denote all subjects or components that are not already

DOI: 10.4324/9781003392545-2

included within STEM (Yakman, 2010). What does the "A" of STEAM mean in Japan, especially in the policies of Japan's government? According to the Ministry of Education, Culture, Sports, Science and Technology (Ministry of Education, Culture, Sports, Science and Technology [MEXT] (Central Council for Education), 2021), this "A" refers to the liberal arts, which encompasses all subjects that are not already included within STEM.

MEXT (2018) published the report "Human Resource Development for Society 5.0," advocating the importance of STEAM education in upper secondary schools and universities. The Cabinet Office (Council on Economic and Fiscal Policy) (2020) also published the policy statement "Basic Policy on Economic and Fiscal Management and Reform 2020," which mandated the reform of elementary and secondary education and support for "gifted and talented individuals [and] STEAM education" (p. 46). Additionally, the Cabinet Office (Council on Science, Technology and Innovation) (2021) published the sixth "Science, Technology, and Innovation (STI) Basic Plan"; in this plan, STEAM education is encouraged in order to develop human resource to support the new "Society 5.0." As these statements suggest, STEAM education is one of the Japanese government's concerns and is linked to human resource development.

Research aim, questions, and methodology

This chapter explores the history of Japan's STEM education. To accomplish this aim, the following research questions were formulated: (1) What is the nature of each STEM subject in the historical context? and (2) Who is STEM literacy for?

To analyze the historical progression of STEM subjects, four periods were delineated, spanning from the mid-19th century to the present day: (1) the mid-19th century to the 1900s (the "take-off" period); (2) the 1910s to the 1940s (wartime; "the drive to maturity" in Japan in 1940); (3) post-WWII decades from the 1950s to the 1980s (the "Japanese economic miracle" era); and (4) the 1990s to the present (the age of STEM education?). In this analysis, I drew upon economic theories, notably those of Rostow (1959, 1960), as economic waves, cycles, or stages are inherently intertwined with scientific and technological innovations. While economic theories face criticism, these innovations and economic growth have influenced the nation's attitude toward education, especially STEM education.

What insights and understandings can a historical approach offer for promoting STEM education? School subjects are historical and social products, meaning that the nature of each subject is socially and historically cultivated. There are challenges in fostering integration and collaborative efforts among STEM subjects without considering their histories and the underlying nature of each subject. Gaining insight into the nature of each STEM subject can, thus, highlight factors and issues related to integration and collaborative initiatives. Carr (1961), a historian, emphasized that it is imperative to have the opportunity to engage in reflective dialogues about the past. Consequently, this chapter seeks to distill valuable historical insights from key periods regarding the essence of each STEM subject, aiming to improve STEM education. This analytical approach also delves into Japan's stance on STEM education and the relationship between economic growth and education.

Theoretical framework

Economic theories

The economist Rostow (1959, 1960) defined the economic stages as follows: (1) the traditional society, (2) the preconditions for take-off, (3) the take-off (Japan: 1878–1900), (4) the drive to maturity (Japan: 1940), and (5) the age of high mass consumption. According to Rostow (1960), "[t]he take-off is the interval when the old blocks and resistances to steady growth are finally overcome" (p. 7), and "[a]fter take-off there follows, then, what might be called the drive to the maturity ... we define it as the period when a society has effectively applied the range of (then) modern technology to the bulk of its resources" (p. 59). In addition, Šmihula (2009) summarized the economic cycles defined by Kondratieff and other economists as follows: (1) the age of coal and steam (1780–1840), (2) the age of railways and mass production (1840–1890), (3) the age of electricity (1890–1940), and (4) the age of electronics and microelectronics (1940–1980). Šmihula (2009) argued that these cycles "must be modified in order to correspond to modern experience" (p. 35). He also described two additional cycles: (5) the age of the information and telecommunications revolution (1985–2015) and (6) the post-informational technological revolution (2015–2035?). Referring to economists' dialogues, I constructed the period classification in this chapter while considering these theories and economic trends.

Education and economic growth

According to Hanushek and Woessmann (2021), a strong relationship exists between cognitive skills, which they described as a nation's knowledge capital, and economic growth; thus, they contended that improving the quality of schooling has long-term benefits for a nation. Moreover, governments such as those of Japan and the U.S. have also noted the relationship between education and economic growth. The Ministry of Education, Science, Sports and Culture (MoE) (1963) attributed the miraculous postwar recovery of the Japanese economy to accumulation resulting from Japan's educational efforts since the Meiji era. The MoE also recommended that the development of scientific creativity, technical competencies, and skills for the workforce would depend on the quantitative and qualitative development of education; in this report, the MoE regarded education as an "investment". In the U.S., Leestma et al. (1987) pointed out some problems in Japan's education system but reported that it contributed substantially to national economic strength. In addition, Johnson (1982) argued that Japan, in modern times, has focused on "nationally supported goal[s] for its economy" (p. 20). Consequently, the economic growth of the nation and education have historically had a close relationship in Japan.

The mid-19th century to the 1900s: the time of "take-off"

According to Rostow (1960), factors during the previous economic cycle serve as preconditions for take-off, such as external intrusion and the establishment of an "effective centralized national state" (p. 7). In Japan's case, the former was Commodore Perry's "black ships" (Watanabe, 1990), and the latter was the Meiji government. The economist Morishima (1982) argued that by "assimilating and absorbing the science of Western Europe with amazing speed" (p. 87), the Meiji government achieved take-off in the economy.

In this process, the Meiji government aimed to provide education *for all* Japanese people, irrespective of their class, creating a modernized nation by learning from Western countries like the U.K., the U.S., and Germany (Prussia). The Department of Education (later known as the MoE and then MEXT) was established in 1871. The department issued the Regulations for the Course of Study for Elementary Schools in 1872. Several mathematical and scientific subjects were introduced from the West, with the allocated lesson hours surprisingly amounting to approximately half of the total lesson time (Itakura, 1968). The emphasis on Western mathematics and science education was, thus, prominent in the early phase of this period. Teaching in middle schools (for boys) placed emphasis on "Western learning and especially upon [the] natural sciences" (MoE, 1980, p. 58). Technology education was categorized as either *shukō* (manual training), an elective, or vocational preparatory education. Although the curriculum at the elementary and middle school levels was influenced by Western models along with separate subject curricula, it was recontextualized for the Japanese cultural and social contexts (Isozaki, 2014). Consequently, the government improved primary education for all, implementing measures to enhance workforce skills and enable effective participation in modern global economic activities.

Regarding higher education, the government hired Western academics as professors at educational institutions, offering them high salaries. Many students also pursued studies abroad in various fields, later becoming professors at higher education institutes. The Imperial College of Engineering (*Kōgakuryō*, later *Kōbu Daigakkō*, and subsequently a part of the Faculty of Engineering at the Imperial University in Tokyo) marked the beginnings of engineering education in Japan and employed Western professors under the instruction of British principal Henry Dyer. Dyer (1904) emphasized the integration of theory and practice, stressing not only specialized engineering education but also general education encompassing mathematics and science as part of foundational learning, along with English and Japanese literature.

During the Meiji era (1868–1912), the Meiji government established the Imperial University in Tokyo, Kyoto, Sendai, and Fukuoka, with these institutions focusing on scientific disciplines such as science, engineering, medicine, and agriculture. Despite the considerable financial burden, the nation continued to invest in training the leading scientists and engineers. Additionally, the Meiji government progressively developed specialized schools for vocational education in engineering at the post-secondary/tertiary level, which eventually became engineering faculties at national universities after WWII. Local governments also established vocational schools that provided engineering/industrial education at the middle/secondary level to train artisans and technicians. These schools included mathematics and science in their general education.

Japan's approach to education, particularly science education, was different from that of the U.K. When Japanese society underwent modernization in the mid-19th century, influenced by the West, science was firmly placed into both the elementary and secondary school curricula, as regulated by the Ministry of Education. In addition, Japan's education system traditionally guaranteed learning opportunities and academic achievement for every child based on laws and regulations. Regardless of gender and social background, *all* students attending schools were required to study science and mathematics. Consequently, with the increase in the enrollment rate in compulsory education prior to WWII (from 45.0% in 1887 to 95.6% in 1905 [MoE, 1980]), and despite secondary education not being compulsory, a policy of "science and mathematics for all" was adopted, albeit imperfectly, by introducing these subjects into Japan's school curriculum. However, the type of "science education" and "mathematics education" provided varied across different types of schools, such as middle schools for boys, girls' high schools, and vocational schools, all of which constituted secondary education. Even students in higher middle school in post-secondary/tertiary education (not at the university level), whether science or non-science majors, were required to study science and mathematics as compulsory subjects.

Consequently, during this period, Japan achieved an economic "take-off" through two key educational measures: the expansion of primary education to elevate the population's educational attainment and higher education that aimed to cultivate the nation's leaders. These developments underscore the Meiji government's emphasis on STEM education as crucial for building a modernized nation and achieving economic growth.

The 1910s to the 1940s: wartime; "drive to maturity" in Japan

Rostow (1960) proposed the concept of the "drive to maturity" as when "a society has effectively applied the range of modern technology to the bulk of its resources" (p. 59) and defined the symbolic date of this drive in Japan as 1940. There were several wars throughout the take-off, the drive to maturity, and the 1940s, including the Russo-Japanese War (1904–1905), World

War I (1914–1918), and WWII (1939–1945). Rapid industrial development increased the demand for workers, ranging from artisans and technicians to professional scientists and engineers. Consequently, scientific, engineering, and vocational education involving industrial education were more in demand than in the previous period. To meet this demand, Japan's government implemented initiatives to enrich the nation by strengthening international economic competitiveness through significant advances in science and technology and reforming the education system.

However, we must not forget that this was a period of war regimes. Consequently, though Japan's capitalism developed rapidly, there was a shift from light to heavy industry. In this period, several educational trends also emerged, such as progressive education; the Perry movement in mathematics education, which emphasized a hands-on approach (Jackson et al., 2020); the heuristic method; the ideas of nature study; and general science in science education, which were influenced by education in the U.K. and the U.S. These trends were introduced and practiced in Japan's schools.

The social changes and new educational trends triggered expanded educational opportunities for the middle social classes and educational reforms. The Special Council for Education was established under the supervision of the prime minister. From the perspective of STEM education, two events warrant discussion in this period: the introduction of the integrated subject "science and mathematics," called *risū-ka* in Japanese, and the *tokubetsu kagaku kyōiku* (Special Science Education: SSE) research project. The Middle Level School Order, which aimed to "train the people for the Imperial Way through higher general education or vocational education" (MoE, 1980, p. 207), was promulgated by the MoE in 1943. Under this order, science and mathematics were first integrated into one compulsory subject area that included three subjects: mathematics, *busshō* (physical sciences), and *seibutsu* (biology). *Busshō* consisted of content on physics, chemistry, earth science, and technology/engineering involving machine maintenance and crafting. *Busshō* can, thus, be seen as one type of STEM education in Japan. However, although collaboration between the three subjects was encouraged, no common learning content was set across them, nor was it taught by their respective specialists.

The beginning of the mobilization of scientists and engineers in Japan was an initiative of "innovative bureaucracy", especially "technocrat bureaucracy" (Hiroshige, 1973, p. 163). Science and engineering universities were expanded with the aim of wartime readiness. In 1943, Japan's government discontinued draft deferments for all students at universities and specialized schools, expecting those studying science, engineering, and technology fields to participate in the war (MoE, 1980). Toward the end of WWII, Japan was defeated, scorched, and experienced food shortages. Under these extremely difficult conditions, the SSE project was established to promote "science for excellence," in contrast to "science for all." The MoE required four higher normal schools—Tokyo, Hiroshima, Kanazawa, and Tokyo Women's Higher

Normal Schools (as well as Kyoto Imperial University, which joined later)—to conduct the SSE research project. These higher normal schools and Kyoto Imperial University organized research committees and used the elementary and secondary schools attached to them or municipal schools to undertake practical research. In 1945, just before the end of the war, the MoE (1945) stated that the aim of the SSE project was "to contribute to the dramatic improvement of science and technology in Japan" (p. 27). Consequently, the school curriculum was enhanced with advanced-level science, mathematics, and technology education aimed at training creative future scientists and engineers who could lead Japan in the world. Just after WWII, in October 1945, Kanazawa Higher Normal School (1945) proposed a new aim for the SSE project: "[T]o improve the lives of Japanese people dramatically and to create a new scientific culture for positively contributing to world peace" (p. 1). The Super Science High Schools (SSH) introduced in 2002 are similar to the SSE project (Isozaki, 2022); thus, this project can be regarded as one of the origins of both STEM education for excellence and science for excellence.

Although they took place in wartime, we should critically examine these two educational events—the introduction of the integrated subject "science and mathematics" and the SSE research project—in relation to the nature of STEM education.

Post-WWII decades from the 1950s to the 1980s: the "Japanese economic miracle" era

After its defeat in WWII, Japan's new government under the General Headquarters of the Allied Powers embarked on postwar reconstruction. The MoE accepted the United States Education Mission to Japan and began to reform education based on American democracy. The Ministry of International Trade and Industry (later the Ministry of Economy, Trade and Industry) was formed in 1949 to promote industry, and the Science and Technology Agency was established in 1956 (and later integrated into MEXT).

The Ministry of Education reorganized the school system and introduced the Course of Study as a national curriculum standard. While mathematics (arithmetic in elementary school) and science were firmly placed in the school curriculum from elementary school through upper secondary school, as in pre-wartime, the positioning of technology education was quite complex.

Technology education was initially positioned as vocational preparatory education in lower secondary schools and later positioned as a general education subject, together with home economics. It eventually became compulsory for both boys and girls. Compared to the early Meiji era described above, the percentage of total class hours for science and mathematics in elementary and lower secondary schools in this period was lower. Figure 1.1 shows that the percentage of classes for mathematics and science generally remained around 20%–30% of the total curriculum hours. *All* students from elementary through upper secondary levels were required to study science and mathematics, even

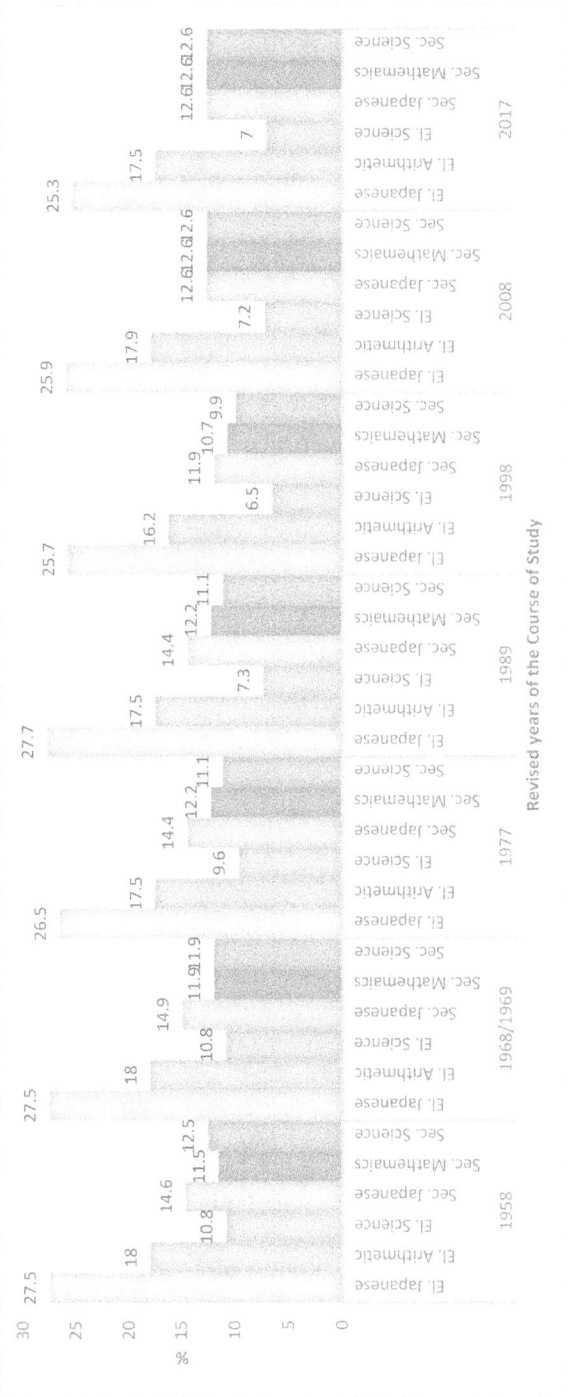

Figure 1.1 Percentage of each subject to total class hours in elementary and lower secondary schools.

Note: Excluding lesson hours for elective subjects for lower secondary schools. In 1989, lower secondary science 10.0–11.1%.

Source: National Institute for Educational Policy Research: https://crid.nier.go.jp/guideline.html, and https://www.nier.go.jp/kiso/sisitu/siryou1/2-02.pdf

non-science students in upper secondary school. *All* students in lower secondary schools were required to take the subject "technology and home economics."

The Course of Study, revised approximately every 10 years, has undoubtedly contributed to guaranteeing the quality of compulsory education and reducing regional disparities in public education. Unlike during wartime, there was no attempt to integrate mathematics and science into the school curriculum as promulgated in the Course of Study during this period, except for the science and mathematics course in the *senmon gakka*—the specialized course at upper secondary schools. However, although the course was named "science and mathematics," scientific subjects and mathematics were taught independently by the teachers of each subject. Japan's government and parliament members as well as the business community regarded education, especially science and technology education, as a vehicle to rebuild a democratic nation and achieve economic growth after WWII. For example, the Industrial Education Promotion Law was enacted in 1951 to promote industrial education, which includes agriculture, commerce, fishery, and other industries. Industrial education was recognized as the basis for developing the industrial economy in Japan. In 1953, the Science Education Promotion Law was also enacted, which regarded science education as a foundation for building a cultured nation. According to Nakayama (1991), "the major target of science and technology promotion was set for economic recovery" (p. 237) in post-wartime in Japan.

Facilitated by the U.S., Japan achieved significant economic growth from the mid-1950s through the first half of the 1970s in what is known as "Japan's economic miracle" (Beckley et al., 2018). Western observers in the 1970s and 1980s determined that education was the key to Japan's successful industrial economic growth; for example, Vogel (1979) claimed that Japan's economic achievement was due in part to the high quality of its compulsory education and the country's guaranteed equality in education.

Why were there no attempts to integrate science, mathematics, and technology into the formal school curriculum during this period? Although there are several possible explanations, the main reason could be that there were no clear or loud voices from the business community and education societies calling for such integration at this time, unlike the period around the millennium. First, as mentioned above, Japan's world status gradually increased because of significant economic growth based on advancements in science and technology. For example, although several economic business organizations, such as *Keidanren*, often issued educational policy statements, this practice increased from the 1990s, at the beginning of the "lost decades" (e.g., Funabashi, 2015). Business organizations published opinions and statements on education from the 1950s onward; however, they primarily focused on human resource development and human abilities to promote industry through science and technology in the 1950s and 1960s. Their statements on education subsequently changed in line with social changes in Japan and its position in the world.

Table 1.1 Japanese students' TIMSS results: rank (no. of countries).

	Elementary arithmetic	Elementary science	Secondary mathematics	Secondary science
1964/1970		1 (16)	2 (12)	1 (18)
1981/1983		1 (19)	1 (20)	2 (26)
1995	3 (26)	2 (26)	3 (41)	3 (41)
1999			5 (38)	4 (38)
2003	3 (25)	3 (25)	5 (45)	6 (45)
2007	4 (36)	4 (36)	5 (48)	3 (48)
2011	5 (50)	4 (50)	5 (42)	4 (42)
2015	5 (49)	3 (47)	5 (39)	2 (39)
2019	5 (58)	4 (58)	4 (39)	3 (39)

Source: International Association for the Evaluation of Educational Achievement: https://www.iea.nl/studies/iea/timss. National Institute for Educational Policy Research: https://www.nier.go.jp/timss/index.html

Second, in international comparisons, Japanese students' science and mathematics scores in this period were the highest in the world (see Table 1.1).

In Japan, the school curriculum from elementary through upper secondary schools was predominantly subject-based, and science and mathematics were tested as independent subjects in entrance examinations for upper-level education, such as upper secondary school and university. Technology education in lower secondary schools, whether included in vocational preparatory or general education or studied by both boys and girls or boys only, was not separated from home economics and was rarely positioned as an entrance examination subject in upper secondary schools.

Third, Japan had been closely monitoring the U.S. and the U.K. and "playing catch up with the West" (Kariya, 2015) since the mid-19th century. However, the West's integration of academic subjects was not a major educational trend that significantly impacted Japan. In the late 1950s through the 1970s, the education movement (*kyōiku no gendaika* in Japanese) in the West strongly influenced the emphasis on science and mathematics education in Japan, especially at the secondary level (e.g., Isozaki, 2021), although this influence only had an effect within each subject and not across subjects. Various educational issues, such as excessive exam competition, school violence, and bullying, became social problems in the 1980s; however, the integration of academic subjects, such as science and mathematics, was not widely discussed during this period.

The 1990s to the present: the age of STEM education?

During the "lost decades," economic growth stagnated for a prolonged period, and Japan's status declined compared to the previous era. As if responding to this economic stagnation, education reform was promoted. Economists Nishimura et al. (2022) critically analyzed the relationship between the lesson hours dedicated to science and mathematics in upper secondary schools for each revision of the Course of Study and the level of research and development (R&D) in science and technology. They concluded that science and mathematics education, fundamental to Japan's economic competitiveness, had a significant impact on its R&D prowess in the long term. Indeed, in the case of lower secondary schools, as shown in the Figure 1.1, the percentages of mathematics and science in the total number of lesson hours have not changed significantly. However, the required number of lesson hours for each subject decreased compared to the two revised Courses of Study in 1969 (mathematics and science each 420 hours) and 1998 (mathematics and science, 315 and 290 hours, respectively). One contributing factor to this was the result of the policy of *yutori kyōiku* (relaxed education).

Since the onset of the lost decades, Japan's government has focused on initiatives in two fields: education and science and technology. Successive prime ministers directly intervened in educational reform, which typically falls under the purview of the Minister of MoE. Kariya (2015) argued that the Japanese government's educational reform in the 1980s and 1990s was based on the notion that Japan should abandon the outdated catch-up model of education. Beginning around the turn of the millennium, educational issues such as *rikagirai* (a dislike of science), *rikabanare* (declining numbers of candidates applying to scientific and engineering faculties at universities), and *gakuryoku-teika* (a crisis due to inadequate academic abilities) became major concerns for the government, business community, and the public. In 1995, the Basic Act on Science and Technology (later, Science, Technology, and Innovation) was enacted to enhance the level of science and technology in Japan. Under this act, Japan's government was mandated to implement policies to advance science and technology learning through formal and informal education (Ministry of Justice, 1995). In 1996, *Keidanren* published an education policy statement (Japan Business Federation (*Keidanren*), 1996). This statement evaluated Japan's "catch-up and pass" model for economic growth: *Keidanren* asserted that this model had run its course and proposed five suggestions and seven action agendas to cultivate creative human resource through education supported by community cooperation.

While education was, to some extent, linked to economic stagnation, were the phenomena of *rikagirai* and *rikabanare*, as well as *gakuryoku-teika*, truly observed? As Tables 1.1 and 1.2 show, Japanese students' academic proficiency in science and mathematics ranked at the highest level globally in the Programme for International Student Assessment and the Trends in International Mathematics and Science Study during this period. Nevertheless, the percentage of lower secondary school students in Japan who expressed an interest in learning science and mathematics or aiming for scientific or mathematical careers in the future consistently remained below the international average in each comparison.

Although the terms *rikagirai* and *rikabanare* denote a dislike of learning science and mathematics and a low percentage of students taking advanced-level science and mathematics subjects in upper secondary schools,

Table 1.2 Japanese students' PISA results: rank (score) 2000–2022.

	Reading literacy	Mathematic literacy	Scientific literacy	Numbers of countries
2000	8 (522)	1 (557)	2 (550)	32
2003	14 (498)	6 (534)	2 (548)	41
2006	15 (498)	10 (523)	6 (531)	57
2009	8 (520)	9 (529)	5 (539)	65
2012	4 (538)	7 (536)	4 (547)	65
2015	8 (516)	5 (532)	2 (538)	70
2018	15 (504)	6 (527)	5 (529)	79
2022	3 (516)	5 (536)	2 (547)	81

Source: National Institute for Educational Policy: https://www.nier.go.jp/kokusai/pisa/.

their essence is an escape from learning. Given these circumstances, MEXT introduced the SSH program in 2002, aiming to enhance creative science and mathematics education in upper secondary schools. As previously mentioned, SSH appears to be a revamped version of the SSE project conducted shortly after WWII and can be regarded as representing science for excellence (e.g., Isozaki, 2022). Additionally, MEXT newly revised the Course of Study for elementary through upper secondary schools in 2017 and 2018. In this revision, a focus on "inquiry" was emphasized in both science and other subjects like history and Japanese in upper secondary schools. A novel subject called "inquiry-based study of science and mathematics" was introduced, and "the period for integrated studies" was restructured as "the period for inquiry-based cross-disciplinary study." Although STEAM education can be provided within subject-specific learning, MEXT encourages it in both the inquiry-based study of science and mathematics and the period for inquiry-based cross-disciplinary study. To effectively deliver this subject and period, teachers must collaborate with their colleagues. The Ministry of Economy, Trade and Industry, which oversees Japan's business community, actively participates in STEAM education, particularly informal education. The Ministry of Agriculture, Forestry and Fisheries is also advancing food and nutrition education as a national movement in Japanese *shokuiku* (see Chapter 7). The outcomes of SSH are anticipated to be reflected in inquiry activities within STEM or STEAM education in upper secondary schools.

Discussion

As pointed out by Jackson et al. (2020), Australia, the U.K., and the U.S. have a long history of integrated STEM curricula. They argued that the Perry movement in mathematics, the inclusion of science in the school curriculum, and progressive education were historical events that led to these integrated STEM curricula. If this is the case, Japan can also be regarded as a country with a long history of an integrated STEM curriculum. The Perry movement and progressive education influenced education in Japan; however, as mentioned above, science and mathematics have been firmly included in Japan's school curriculum since the mid-19th century. The importance of STEM fields in Japan does not begin with the present time but was called for by modernization, two world wars, postwar reconstruction, and global economic competition based on utilizing science and technology. As mentioned above, the importance of education for economic growth is also observed by economists; for example, Barro (2013) pointed out that the quality and quantity of a nation's schooling are related to its economic growth and that scores on science tests have a particularly "strong positive relationship with economic growth" (p. 326).

Tracing back to WWII, we can see the emergence of an integrated STEM curriculum across subjects such as science, mathematics, and technology in Japan. Historically, while STEM education was enacted as "SteM" in the general education course in upper secondary schools due to the lack of technology and engineering as a separate subject, it was enacted as "sTEm" in vocational schools, where science and mathematics were basic subjects as general education that supported students' learning of technology and engineering. This situation applies not only in Japan but also in other countries. Recognizing the history of STEM subjects may show that all four subjects do *not* need to be represented equally within the school curriculum, given their nature. However, as both future STEM-literate citizens and future scientists, students should be able to effectively identify the importance of the connection between these subjects to enhance their abilities and apply their learning to relevant real-world problems in social and daily life contexts.

There are two ideas on STEM education. On the one hand, some advocates stress a nation's need for an essential workforce resource, such as future scientists, engineers and technicians, to help the country excel in international economic competition through science and technology. On the other hand, others argue that we should prepare all students to be STEM-literate future citizens who can conduct inquiry activities and solve real-world problems by applying the knowledge and skills learned in STEM subjects. The former idea represents "STEM for excellence," while the latter represents "STEM for all."

STEM literacy is generally defined as the goal of STEM education. When we consider STEM literacy, the definition of scientific literacy is helpful. Roberts (2007) and Roberts and Bybee (2014) suggested two visions of scientific literacy: Vision I and Vision II. In Vision I, scientific literacy involves an individual acquiring scientific concepts and knowledge and using the scientific ways of thinking that they have learned to solve problems. In Vision II, scientific literacy involves applying scientific ideas and practices to real-world problems in social and daily life contexts. Many new global issues are complex, unpredictable, and interdisciplinary and cannot be solved by one academic discipline, such as science, alone. As Fisher and Frey (2014/15) stated, "[a]ll students—whether or not they pursue careers in science, technology, engineering, and math—will be consumers of news and information on STEM issues that will directly affect their lives" (p. 86). Thus, as future STEM-literate citizens, all students will be required to make evidence-based, rather than ideological, decisions on such issues. Zollman (2012) highlighted that although there is no consensus among professional organizations on STEM literacy, most definitions include societal and economic needs but overlook personal needs. As Japan's history of education exemplifies, the nation's needs for economic growth have been emphasized over benefitting individual needs in education. Therefore, STEM literacy should include the personal aspect of making decisions, for example, on global issues. If we are conceptualizing STEM literacy, Vision II scientific literacy should adopt "STEM literacy for all."

STEM literacy in Japan must be defined by monitoring global trends and then recontextualizing and understanding the historical nature of each subject, rather than simply accepting foreign definitions. Furthermore, STEM literacy is best defined from the perspective of the three competencies: knowledge and skills; the ability to think, make judgments, and express oneself; and the motivation to learn and humanity. Japan's Course of Study currently requires these three competencies to be cultivated. Finally, considering the aforementioned points, rather than viewing "STEM for all" and "STEM for excellence" as a distinct dichotomy, viewing "STEM for excellence" as encompassed by "STEM for all" should be encouraged.

Conclusion

The historical approach taken in this chapter allows readers to understand the nature of each STEM subject within the social and economic contexts of Japan. Understanding the history of STEM education is an essential way to create a new perspective on it.

From a historical perspective, Isozaki and Isozaki (2021) reflected on the fact that the movement for science and technology in Japanese society has remained in the theoretical realm and has not extended as far as practice in schools. They argued that to create Japanese-style STEM education, we should consider the following points: (1) the aims and objectives of STEM (from the perspectives of STEM for all and STEM for excellence); (2) the significance and values of STEM subjects; (3) collaboration among teachers in STEM subjects; and (4) the meaning of learning STEM for teachers, learners, and their parents. Historically, the success or failure of education has depended more on the enthusiasm and quality of teachers than legal arrangements and budgetary measures. In both Japan and other countries, the school curriculum has traditionally been subject-based; therefore, teachers are not sufficiently familiar with their colleagues' work. To provide effective STEM education, teachers must carefully consider the aforementioned points and collaborate with their colleagues.

Though a critical review of history cannot provide a complete vision of the future, insights from historical research like those presented in this chapter can suggest a potential future direction for Japan's STEM education.

Acknowledgments

I would like to extend my special and heartfelt thanks to Jinwoong Song, professor at Seoul National University in South Korea, for his valuable insights.

This work was supported by the Japan Society for the Promotion of Science KAKENHI Grant Numbers JP20K20832, JP21H00919 (JP23K20744), JP22K02982.

References

Barro, R. J. (2013). Education and economic growth. *Annals of Economics and Finance, 14*, 301–328.
Beckley, M., Horiuchi, Y., & Miller, J. M. (2018). America's role in the making of Japan's economic miracle. *Journal of East Asian Studies, 18*(1), 1–21.
Breiner, J. M., Johnson, C. C., Harkness, S. S., & Koehler, C. M. (2012). What is STEM? A discussion about conceptions of STEM in education and partnership. *School Science and Mathematics, 112*(1), 3–11.
Bybee, R. W. (2010). What is STEM education? *Science, 329*(5995), 996. https://doi.org/10.1126/science.1194998
Cabinet Office (Council on Economic and Fiscal Policy). (2020). *Basic policy on economic and fiscal management and reform 2020: Overcoming the crisis and moving toward a new future.* Cabinet Office. Retrieved from https://www5.cao.go.jp/keizai-shimon/kaigi/cabinet/honebuto/2020/2020_basicpolicies_en.pdf
Cabinet Office (Council on Science, Technology and Innovation). (2021). *Science, technology, and innovation basic plan.* Cabinet Office. Retrieved from https://www8.cao.go.jp/cstp/english/sti_basic_plan.pdf
Carr, E. H. (1961). *What is history? The George Macaulay Trevelyan lectures delivered in the University of Cambridge January-March 1961.* Macmillan.
Committee on STEM Education of the National Council of Science & Technology Council. (2018). *Charting a course of success: America's strategy for STEM education.* The White House. Retrieved from https://files.eric.ed.gov/fulltext/ED590474.pdf
Dyer, H. (1904). *Dai Nippon: The Britain of the East: A study in national evolution.* Blackie & Son.
Fisher, D., & Frey, N. (2014/15). STEM for citizenship. *Educational Leadership, 72*(4), 86–87.
Funabashi, Y. (2015). Introduction. In Y. Funabashi & B. Kushner (Eds.), *Examining Japan's lost decades* (pp. xx–xxxiv). Routledge.
Hanushek, E. A., & Woessmann, L. (2021). Education and economic growth. In *Oxford Research Encyclopedia of Economics and Finance* (pp. 1–23). Oxford University Press. Retrieved from http://hanushek.stanford.edu/site/default/files/publications/Hanushek%2BWoessmann%202021%20OxfResEncEcoFin.pdf (https://doi.org/10.1093/acrefore/9780190625979.013.651)
Hiroshige, T. (1973). *Kagaku no shakaishi* [*The social history of science*]. Chūōkōronsha [in Japanese].
Isozaki, T. (2014). The organisation and the recontextualization of *rika* (school science) education in the second half of the nineteenth century in Japan. *Science & Education, 23*(5), 1153–1168. https://doi.org/10.1007/s11191-013-9615-4

Isozaki, T. (2021). Why research the history of science education/teaching (rika) in Japan? In T. Isozaki & M. Sumida (Eds.), *Science education research and practice from Japan*(pp.1–23).SpringerNature.https://doi.org/10.1007/978-981-16-2746-0_1

Isozaki, T. (2022). A historical perspective of science education in Japan: Which way is it headed in the future? *Asia Pacific Journal of Educators and Education, 37*(2), 167–184. https://doi.org/10.21315/apjee2022.37.2.8

Isozaki, T., & Isozaki, T. (2021). Nihon-gata STEM kyōiku no kouchiku ni mukete no rironteki-kenkyū–Hikaku-kyōikugakuteki-shiza kara no bunseki wo tōshite (Theoretical research for establishing a Japanese-style STEM education: Analysis for a comparative historical point of view). *Journal of Science Education in Japan, 45*(2), 142–154 [in Japanese].

Itakura, K. (1968). *Nihon-rikakyōikushi–fu nenpyō (A history of science education in Japan with a chronological table).* Daiichihouki [in Japanese].

Jackson, C., Tank, K. M., Appelgate, M. H., Jurgenson, K., Delaney, A., & Erden, C. (2020). History of integrated STEM curriculum. In C. Johnson, M. J. Mohr-Schroeder, T. J. Moore, & L. D. English (Eds.), *Handbook of research on STEM education* (pp. 169–183). Routledge.

Japan Business Federation (*Keidanren*). (1996). Developing Japan's creative human resources: An action agenda for reform in education and corporate conduct. *Keidanren.* Retrieved from http://www.keidanren.or.jp/english/policy/pol043.html

Johnson, C. (1982). *MITI and Japanese miracle: The growth of industrial policy, 1925–1975.* Stanford University Press.

Kanazawa Higher Normal School. (1945). Tokubetsu-kagakukyōiku-jisshi-yōkō-an (The proposal implementation for special science education (Draft): dated on the 11th October 1945) (pp. 1–12) [Unpublished]. *Kagaku kankei shiryō (Documents on Science Education (Miscellaneous)).* Kanazawa, Japan [in Japanese].

Kariya, T. (2015). The two decades in education: The failure of reform. In Y. Funabashi & B. Kushner (Eds.), *Examining Japan's lost decades* (pp. 101–117). Routledge.

Leestma, R., August, R. L., George, B., Peak, L., Shimahara, N., Cummings, W. K., Stacey, N. G., Bennett, W. J., & Dorfman, C. H. (1987). *Japanese Education Today. A report from the U.S. Study of Education in Japan.* Government Printing Office, Washington, D.C. Retrieved from https://files.eric.ed.gov/fulltext/ED275620.pdf

Ministry of Education, Science, Sports and Culture (MoE). (1945). *Tokubetsu kagaku kyōiku jisshi ni kakawaruken* (Research on the implementation of special science education). *Monbu-Jiho, 822,* 27 [in Japanese].

Ministry of Education, Science, Sports and Culture (MoE). (1963). *Japan's growth and education: Educational development in relation to socio-economic growth.* MoE.

Ministry of Education, Science, Sports and Culture (MoE). (1980). *Japan's modern educational system: A history of the first hundred years.* Ministry of Finance.

Ministry of Education, Culture, Sports, Science and Technology (MEXT). (2018). *Society 5.0 ni muketa jinzaiikusei~Shakai ga kawaru, manabi ga kawaru~ (Human resource development for Society 5.0: Changes to society, changes to learning)* [in Japanese]. Retrieved from https://www.mext.go.jp/component/a_menu/other/detail/__icsFiles/afieldfile/2018/06/06/1405844_002.pdf

Ministry of Education, Culture, Sports, Science and Technology (MEXT) (Central Council for Education). (2021). *"Reiwa no nihongata-gakkōkyōiku" no kouchiku wo mezashite: Subete no kodomotachi no kanousei wo hikidasu, kobetsusaitekinamnabi to kyōdoutekina manabi no jitugen (tōshin) (Toward the construction of "Japanese style school education": Optimal individualized learning and collaborative learning that bring out the potential all children (report)* [in Japanese]. Retrieved from https://www.mext.go.jp/content/20210126-mxt_syoto02-000012321_2-4.pdf

Ministry of Justice (1995). *Basic act on science and technology.* (Act No. 130 of November 15, 1995) by Japanese Law Translation [in Japanese and English]. Retrieved from https://www.japaneselawtranslation.go.jp/ja/laws/view/2761

Morishima, M. (1982). *Why has Japan succeeded? Western technology and the Japanese ethos.* Cambridge University Press.

Nakayama, S. (1991). *Science, technology and society in postwar Japan.* Kegan Paul International.

Nishimura, K., Miyamoto, D., & Yagi, T. (2022). Japan's R&D capabilities have been decimated by reduced class hours for science and math subjects. *Humanities and Social Sciences Communications, 9,* 1–9. https://doi.org/10.1057/s41599-022-01234-0

Roberts, D. A. (2007). Scientific literacy/science literacy. In S. K. Abell & N. G. Lederman (Eds.), *Handbook of research on science education* (pp. 729–780). Lawrence Erlbaum Associates.

Roberts, D. A., & Bybee, R. W. (2014). Scientific literacy, science literacy, and science education. In N. G. Lederman & S. K. Abell (Eds.), *Handbook of research on science education* (Vol. II, pp. 545–558). Routledge.

Roberts, G. (2002). *SET for success: The supply of people with science, technology, engineering and mathematics skills (The report of Sir Gareth Roberts' Review).* HM Treasury. Retrieved from https://www.education-uk.org/documents/pdfs/2002-roberts-review.pdf

Rostow, W. W. (1959). The stages of economic growth. *The Economic History Review, 12*(1), 1–16.

Rostow, W. W. (1960). *The stages of economic growth: A non-communist manifesto.* Cambridge University Press.

Šmihula, D. (2009). The waves of the technological innovations of the modern age and the present crisis as the end of the wave of the informational technological revolution. *Studia Politica Slovaca, 2*(1), 32–47.

U.S. Department of Commerce, Economics, and Statistics Administration. (2011). *STEM: Good jobs now and for the future.* U.S. Department of Commerce, Economics, and Statistics Administration. Retrieved from https://www.commerce.gov/sites/default/files/migrated/reports/stemfinalyjuly14_1.pdf.

Vogel, E. F. (1979). *Japan as number one: Lessons for America.* Harvard University Press.

Watanabe, M. (1990). *The Japanese and Western science* (O. T. Benfey, Trans.). University of Pennsylvania Press.

Yakman, G. (2010). What is the point of STE@ M? – A brief overview. STEAM: A framework for teaching across the disciplines. *STEAM Education, 7,* 1–9. Retrieved from https://www.researchgate.net/profile/Georgette-Yakman-2/publication/327449281_What_is_the_point_of_STEAM-A_Brief_Overview/links/5b901b98a6fdcce8a4c2f290/What-is-the-point-of-STEAM-A-Brief-Overview.pdf

Zollman, A. (2012). Learning for STEM literacy: STEM literacy for learning. *School Science and Mathematics, 112*(1), 12–19. https://doi.org/10.1111/j.1949-8594.2012.00101.x

2 Conceptual framework from STEM to STEAM

Through the promotion of STEAM education in Japan

Tadashi Ohtani

Introduction

The development of electricity and machinery since the Industrial Revolution has significantly changed our lives and society. Transportation-related technologies such as automobiles and airplanes have made our movements convenient, while technologies like washing machines, vacuum cleaners, and telephones have made our lives more comfortable. Additionally, in recent years, information processing and communication technologies have expanded the network of information and communication between our lives and society. Information technology also has been integrated with conventional technologies, such as artificial intelligence. Our quality of life has been significantly improved by all of these technologies. In light of this advancement, the development of science and technology and the application of the theory of mathematics, which forms the basis of science and technology, are significant.

The relationship between science, technology, and mathematics is discussed in this chapter, as such a relationship has contributed to the development of science and technology in such a society. Additionally, the spread and promotion of STEM/STEAM education have expanded globally in recent years as education related to science, technology, and mathematics. The spread of STEM/STEAM education in Japan is also discussed in this chapter. In addition, as STEAM education spreads and is promoted in Japan, cross-curricular learning in Japanese school education is also discussed in relation to STEAM education.

How should we understand STEM education in the United States?

The United States is the birthplace of STEM/STEAM education, promoting the concept of "Science for All Americans," which is based on the background of creating, shaping, and responding to major changes in life and society (The American Association for the Advancement of Science [AAAS], 1990). The

DOI: 10.4324/9781003392545-3

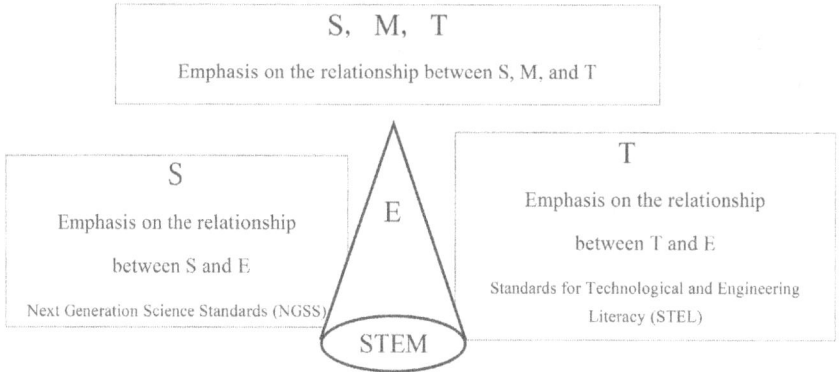

Figure 2.1 Relationship between STEM education policies in the United States.

science (S), technology (T), and mathematics (M) education outlined in this report is indispensable for today's children, who will be in charge of determining the course of the world's future. This is the core concept of "S," "T," and "M" in STEM education.

Figure 2.1 shows a diagram of the relationship between representative educational policies related to STEM in the United States. In addition to the importance of science (S), mathematics (M), and technology (T) education noted by the AAAS, in 2010, STEM education came to be emphasized, and a new standard for science education that emphasizes the connection with engineering (E) was presented by the National Research Council in the United States (National Research Council [NRC], 2013). In line with this, the International Technology Education Association changed its name to the academic society of the International Technology and Engineering Education Association (ITEEA) in 2006; since then, the ITEEA has established a standard for technology and engineering education that emphasizes the relationship between "T" and "E" (International Technology and Engineering Educators Association [ITEEA], 2020).

In various fields in the United States, educational measures related to "S," "M," and "T" are being encouraged. Figure 2.1 shows the relationship between "S," "M," and "T" with engineering (E) that has been expressed in the form of STEM. The direction of "S," "M," and "T" education required in the new era and trends in STEM human resources have led to the development of the STEM education, in which science (S), mathematics (M), and technology (T) are being related to the activities of engineering (E). We must therefore take into account the educational elements found in the engineering (E) activities that will be necessary in the future and have influenced our society using S, M, and T in order to understand the STEM education employed in the United States. Extracting these elements leads to the constructive concept of STEM/STEAM education.

Trends in STEM/STEAM education in Japan

Since 2010, activities based on STEM education have been promoted in Japan in several areas, including schools, private educational institutions, and play-grounds. Students are taught IT technologies (e.g., programming) in STEM education at private educational institutions through activities like program-ming classes. These activities not only teach IT skills but also foster necessary social abilities such as initiative, creativity, and problem-solving ability.

The Science and Technology Basic Plan was developed from a long-term perspective based on the Basic Act on Science and Technology enacted by the Cabinet Office in 1995, in response to the advantages of STEM education and in connection with human resource development and education related to science and technology in Japan (Ministry of Justice, 1995). Since 1996, this plan has been developed once every 5 years, and the 6th Science and Technol-ogy Basic Plan is currently being formulated. The 5th Science and Technology Basic Plan, formulated in 2016, stated recommended advancing student edu-cation to increase their interest and knowledge in science and technology. In Japan, a policy specific to science and mathematics education is indicated for STEM education (Government of Japan, 2016). However, the 6th plan, which was formulated in 2021, had a strong emphasis on improving problem-finding and problem-solving skills as part of the STEAM education from the elementary to the secondary education levels (Government of Japan, 2021).

In 2018, the Ministry of Economy, Trade, and Industry (METI) launched a study on the "Future Classrooms" as a vision for school education and cor-porate training in Japan (Ministry of Economy, Trade and Industry [METI], 2023). The following are the three pillars of "Future Classroom": STEAM learning; self-reliance and individual optimization of learning; and creation of a new learning infrastructure. Figure 2.2 demonstrates how STEAM-based learning stimulates curiosity in each individual, enables the acquisition of

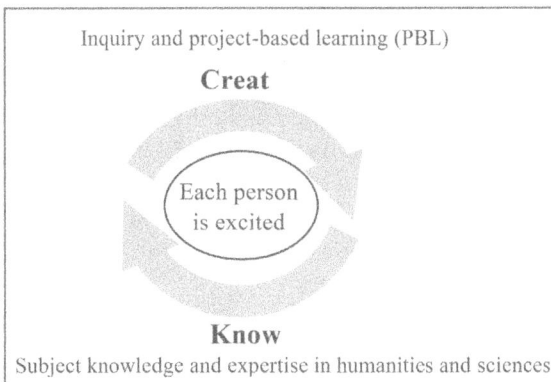

Inquiry and project-based learning (PBL)

Creat

Each person
is excited

Know

Subject knowledge and expertise in humanities and sciences

Figure 2.2 Approach to STEAM learning by METI.

subject knowledge and specialized knowledge regardless of the humanities and sciences, and fosters inquiry/project-based learning. Moreover, it proposes cyclical learning of thinking creatively and logically and identifying unknown problems and their solutions. In recent years, METI has developed a STEAM library with the digital contents to achieve STEAM learning. In this library, practical teaching techniques for learning, like providing worksheets and teaching plans, are provided.

In accordance with such a proposal, METI emphasized that the Japan's education industry lags behind other advanced nations against the current situation wherein various services are expanding in the global education market. In order to integrate technology into the future classroom and education, METI established the Education Industry Service Office in 2017 and launched the "Ed+Tech" study group. The promotion of STEAM education by METI is reflected in the 6th Science and Technology Basic Plan; subsequently, STEAM education is being promoted in Japan.

In terms of the characteristics of STEAM learning, STEAM education serves as a cross-curricular tool, and as the system of learning in school education, it may be able to replace the traditional disciplinary education. However, the new education using digital technology in Japan is lagging behind. Through STEAM learning, students may be able to have the ability to solve their own problems ("creative problem discovery and solution ability"), envision the future, and develop the ability to generate effective solutions. Notably, STEAM learning incorporates art (A) to foster creativity in addition to strengthening the educational field.

The reforms of domestic educational services have been promoted by information technology, according to the attempts of METI as described above, which recognize educational services as an industry. Based on the concept of the fourth Industrial Revolution (Klaus, 2016), the new education created by such information technology is anticipated to contribute to solving various social issues, including providing new jobs and services in Japan. This way of thinking is similar to that mentioned by Bybee in "The Case for STEM Education." Based on this previous study, STEM education differs from previous educational reforms because it addresses global issues, recognizes environmental problems, and develops new workforce skills (Bybee, 2013). Additionally, the METI's attempt to convert learning into STEAM exerts efforts that even extend to promoting specific teaching methods and the understanding of how the creative education should be based on STEM has gained a growing interest. Consequently, the policies are shifting to encourage STEAM education.

Framework for cross-curricular learning in Japanese school education

In order to achieve cross-curricular learning and free instruction by teachers in Japanese school education, the establishment of an integrated study period has been proposed. The "period for integrated studies" has been conducted in elementary and junior high schools since 2002, as well as in senior high schools since 2003. In the period for integrated study, it is important to conduct interdisciplinary and comprehensive study that transcends subject boundaries, as well as to engage in exploratory and collaborative learning. By studying only traditionally established subjects, the integrated study period may help to foster learning for the needs of the modern world. However, because schools differ in decisions and specific teaching contents based on the actual situation, the actual study for the "period for integrated studies" does not show detailed teaching content throughout the study (Tomizawa, 2020).

In terms of the method of learning in the "period for integrated studies," the process of inquiry, as depicted in Figure 2.3, is indicated in the course of study. The "problem setting" in the figure suggests that students should identify the problems by themselves according to the questions and concerns that emerge in both their daily lives and society. The "collecting information" implies that information about specific problems should be collected, and the "organize/analyze" suggests that students should organize and analyze information linked to knowledge and skills and exchange ideas to solve problems. As for the "summary/expression," students should summarize and express the clarified ideas and viewpoints, determine new problems from there, and then begin solving further problems in an expansive manner.

This type of exploratory learning is characterized by learning that seeks and ascertains the essence of things. Its purpose is to utilize knowledge and skills of students that are crossed in learning, to broaden their perspective on learning, and to boost their motivation for further learning. Emphasis is placed on exploratory learning, whereby students consider what they have learned in relation to themselves, become aware of their own growth, and consider their own way of life. This type of interdisciplinary learning exemplifies thinking based on Dewey's empiricism, that is, people's own conception changes as they pursue and deepen their own understanding.

Since the 2000s, new efforts have been made in the Japanese "period for integrated studies" as described above. From the perspective of STEM/STEAM education, themes related to information technology utilization are partially incorporated in the period for integrated studies, and themes related to environmental education are introduced from a scientific perspective. However, the METI-proposed new learning, such as the current learning using digital

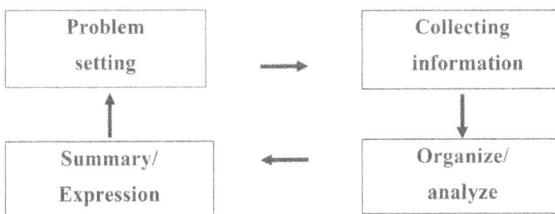

Figure 2.3 Learning method in the "period for integrated studies."

technology, has not been fully implemented into cross-curricular learning. Additionally, the learning is not expanded to include tasks like creating prototypes that are actually based on design thinking in terms of the content of engineering, which is at the core of STEM/STEAM education. In response to this situation, STEAM education enhancement has been promoted in the "period for integrated studies" in high schools to include the advanced contents that can be helpful in developing problem identification and solving skills of students.

Concept of STEAM education based on the Japanese subject framework

Japanese education has encouraged STEAM learning as indicated by METI to foster students' ability to creatively identify problems, solve them, and come up with effective solutions. It intends not only to advance the STEM education field but also to incorporate "A" from the STEAM education to enhance students' creativity in discovering problems. To promote these innovative problem-finding and problem-solving skills in school education, a cross-curricular approach based on the school education system in each country should be established.

Figure 2.4 shows the structural concept of STEAM education based on the Japanese subject framework. Education based on the current industry situation in Japan should be made available for the creative problem discovery and solution. We previously analyzed the qualities and abilities of innovative

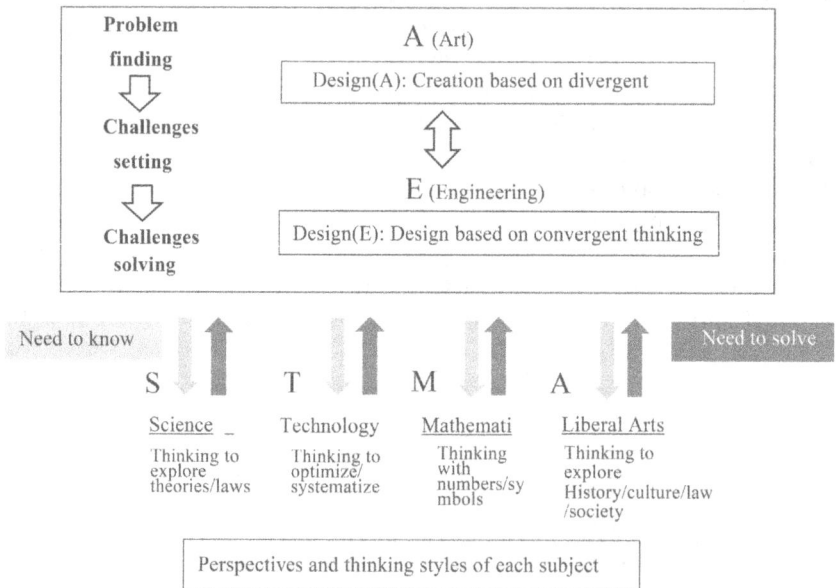

Figure 2.4 Concept of STEAM education based on the Japanese subject framework.

human resources from the perspective of the industry, government, and academia (Iida & Ohtani, 2015), and we found that the ability to discover issues and solve challenges is important. As shown in Figure 2.4, in addition to the ability of "setting of challenges" derived from the findings of our research, the problem-solving process involving "discovery of issues" and "solving of challenges" is necessary in the construction of the concept of STEAM education. Additionally, when discussing from the perspective of the industry, government, and academia, the notation of "issues discovery" is recognized as "issues." Notably, "problem discovery" is written in STEAM education from the perspective of the learner, and it is a "problem" to identify as one's own problem. In addition, in contrast to the learning tasks used in the school education field, the "problem discovery" and "setting of challenges" here mean to lead the global issues and environmental problems, indicating the social issues faced by people worldwide.

Furthermore, A should be added to STEM in the problem-solving process of STEAM education so that we can work on creative problem discovery and solution. The nature of creation requires Design (A) activities that involve more diffuse design thinking. These activities are comparable to those put forth by Jon Maeda, who proposed the need for STEM+A in the United States (Maeda, 2013). However, given that the activities must ultimately converge to an effective solution, the creative activities accompanied by this convergent thinking are performed through Design (E) activities. The creation of an ideal form in accordance with a convergent purpose (problem setting) that is set to design for someone is referred to as a Design (E) activity. This ideal form corresponds to the problem-solving activities in STEM education described by Bybee (2010). Therefore, the activities of Design (A+E) shown in Figure 2.4 are necessary in order to discover and solve problems creatively through the STEAM learning indicated by METI. The Design (A+E) activities overlap one another and go back and forth.

Additionally, the Design (A+E) activities in STEAM education have promoted the development of creative discovery of problems, setting of challenges, and solving of challenges. However, in order to create new knowledge, students should pause and reflect on the creative activities as well as the inquiry activities. In this research activity, by observing phenomena, it can be thought to explore theories/laws for "S," to optimize/systematize for "T," to think with numbers/symbols for "M," and to explore the history, culture, current law and society, and more for "A." In the concept of Design (A), "A" here adds the idea of Liberal Arts as described by Yakman (2008), and the idea of developing Liberal Arts from a new perspective of inquiry can be presented.

Therefore, these activities complement each other through their relationship between the "need to do" activities carried out through design and the "need to know" activities carried out through exploration. The relationships between STEM, as previously proposed by Yata et al. (2020), and STEAM can be expanded in the concept of STEAM education based on the Japanese subject framework.

Learning and teaching through inquiry and creation in STEAM education

As mentioned previously, the creative design activities are important for promoting STEAM education that creatively discovers and solves problems and generates effective solutions. However, these activities need the power of inquiry. It is important to explore the phenomena necessary to design based on divergent thinking, as well as the elements necessary to design in detail based on convergent thinking. Thus, the mutual complementarity of design and inquiry can be used to describe the activities in STEAM education. For STEAM learning to be effective, it must be able to go back and forth between design and inquiry.

The addition of the back-and-forth model of design and inquiry is shown in Figure 2.5, which is proposed by Kolodner by using the concept of Learning by Design (Kolodner, 2002). As can be seen in the figure, a learning model can be applied in the back-and-forth relationship between the "design/redesign" and "investigate/explore," which is the process of Design (A+E) activities for "need to know" and the inquiry activities of "S," "T," "M," and "A (Liberal Arts)" for "need to do." Figure 2.5 displays the reciprocal relationship between the activity of creation through the cycle of design/redesign (left side) and the activity of inquiry through the cycle of investigate/explore (right side). In other words, the design activity (left side) is the activity that utilizes what was learned through the inquiry activity (right side), while the inquiry activity (right side) shows the mutual complementation that arises from the inquiry problems induced by "design/redesign." By utilizing what has been learned in each subject and reciprocating the design and inquiry processes, effective interdisciplinary and comprehensive learning may be developed in STEAM education if such mutually complementary activities are applied.

Regarding the instruction for learning shown in Figure 2.5, the inquiry learning (right side) is typically conducted in the curriculum of science,

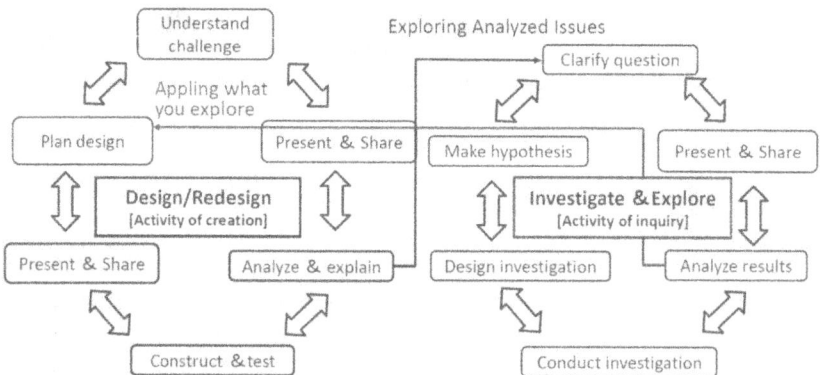

Figure 2.5 Cycle of exploration and design in STEAM education.

mathematics, etc. On the other hand, the study of design and redesign (left side) is promoted in the design study of technology courses in junior high schools. Teachers in charge of each subject implement these teaching strategies. The nature of inquiry-based learning, however, prevents teachers who specialize in it from developing the creative learning methods right away. To put it simply, they are being told unreasonably, "I became a science teacher because I like science, which explores phenomena hidden in nature, but please teach me techniques that design a new value." To prevent such a situation, the Japanese education system provides training for teachers according to their area of specialization. Therefore, by assigning coordinators and supporting the staff in promoting team teaching by multiple teachers and STEAM education would help implement effective STEAM education in a systematic manner.

Conclusion

This chapter provides an overview of STEM education policies in the United States and considers the relationship between science, mathematics, technology, and engineering in STEM education. Based on the STEM concept in the United States, this chapter highlights how discussions in Japan are progressing from STEM education to STEAM education, starting with the concept of STEAM developed by the Ministry of Economy, Trade and Industry. STEAM education here has been developed into STEAM by incorporating the idea of art (A) into creation, and furthermore, the Ministry of Education, Culture, Sports, Science and Technology's approach to STEAM education has included liberal arts (A) in addition to art. The approach was featured by the cross-curricular features that incorporate the integrated concepts of arts and sciences.

Furthermore, as a cross-sectional subject that puts STEAM education into practice, this chapter focused on the comprehensive learning of the "period for integrated studies" in Japanese school education and addressed the features of the learning process during the "period for integrated studies." The framework of STEAM education for developing cross-curricular problem-solving learning was discussed in light of these characteristics as well as the structure of subjects in Japanese school education. Furthermore, the discussion was explained from the importance of the characteristic of STEAM education, which goes back and forth between learning methods based on "inquiry (investigation/explore)" and creation (design).

References

Bybee, R. (2013). *The case for STEM education: Challenges and opportunities* (pp. XI–XII). National Science Teachers Association Press.
Bybee, R. W. (2010). What is STEM education? *Science, 329*(5995), 996.
Government of Japan. (2016). *Kagakugijyutu kihon keikaku (Science, technology, and innovation basic plan)* (pp. 24–34) [in Japanese]. Retrieved from https://www8.cao.go.jp/cstp/kihonkeikaku/5honbun.pdf

Government of Japan. (2021). *Science, technology, and innovation basic plan* (pp. 74–81). Retrieved from https://www8.cao.go.jp/cstp/english/sti_basic_plan.pdf

Iida, R., & Ohtani, T. (2015). Kagakugijyutsu inovasyon no jinzai ikusei no siten kara mita gijyutsuka kyouiku de ikuseisareru sisitu nouryoku no bunseki [Analysis of learning ability in junior high school technology education from the view of development of human resources for science and technology innovation]. *Journal of Science Education in Japan, 39*(2), 104–113 [in Japanese]. Retrieved from https://doi.org/10.14935/jssej.39.104

International Technology and Engineering Educators Association (ITEEA). (2020). *Standards for technological and engineering literacy: The role of technology and engineering in STEM education.* Retrieved from https://www.iteea.org/stel

Klaus, S. (2016). *The fourth industrial revolution* (pp. 6–9). Crown Business.

Kolodner, J. L. (2002). Learning by Design™: Iterations of design challenges for better learning of science skills. *Cognitive Studies, 9*(3), 338–350.

Maeda, J. (2013). STEM + Art = STEAM. *The STEAM Journal, 1*(1), 1–3.

Ministry of Economy, Trade and Industry (METI). (2018). *Mirai no Kyositsu Learning (Innovation – The classroom of the future)* [in Japanese]. Retrieved from https://www.learning-innovation.go.jp/about/

Ministry of Justice. (1995). *Basic act on science and technology.* (Act No. 130 of November 15, 1995) by Japanese Law Translation [in Japanese and English]. Retrieved from https://www.japaneselawtranslation.go.jp/ja/laws/view/2761

National Research Council (NRC). (2013). *Next generation science standards: For states, by states.* National Academy Press.

The American Association for the Advancement of Science (AAAS). (1990). *Science for all Americans* (pp. 25–38). Oxford University Press.

Tomizawa, M. (2020). How integrated study became expansive learning in Japanese elementary schools: The three dimensions of expansion. *An International Journal of Human Activity Theory, 4*, 37–52.

Yakman, G. (2008). *STΣ@M education: An overview of creating a model of integrative education.* Retrieved from https://www.researchgate.net/publication/327351326_STEAM_Education_an_overview_of_creating_a_model_of_integrative_education

Yata, C., Ohtani, T., & Isobe, M. (2020). Conceptual framework of STEM based on Japanese subject principles. *International Journal of STEM Education, 7*(1), 1–10.

3 What are the student competencies that STEM education cultivates?

Based on a comparison between Japan and Germany

Yusuke Endo

Introduction

Over the past 20 years, considerable attention has been paid to the research and practice of STEM education. Although the term STEM is said to have originated in the United States (Sanders, 2009, p. 20), it has spread not only to English-speaking countries but also to non-English-speaking countries worldwide, making it a global trend.

Nevertheless, there are different ways of perceiving what STEM (education) means, and there is not always a common understanding (e.g., Boon Ng, 2019; Isozaki & Isozaki, 2021, p. 143; van Driel et al., 2018, p. 32). There has been a great deal of STEM-related literature that includes sections such as "What is STEM?" and "What is STEM education?" There have been many discussions about the definition and conceptualization of STEM, and it seems that there are several perspectives that have led to diverse interpretations. One, for example, is a perspective on the range of disciplines (subjects) that comprise STEM. As is well known, STEM is an acronym for science, technology, engineering, and mathematics, but the term sometimes includes health sciences, medicine, and agriculture (Tytler, 2020, p. 22), and there is a recent movement to explicitly expand the scope to "STEAM" by adding "Arts" (e.g., Khine & Areepattamannil, 2019). In Germany, MINT (*Mathematik, Informatik, Naturwissenschaften, Technik*) is the established equivalent term of STEM (Tönnsen, 2022, p. 100), in which *Informatik* (computer science) is clearly positioned. As such, there are some differences in terms of which disciplines are included. Another is the perspective on the relationship between the disciplines (subjects) that comprise STEM, for example, whether STEM should be perceived solely as four separate disciplines or as an integration of multiple disciplines. In this regard, Bybee (2013, pp. 73–80) presents nine typologies of how STEM can be perceived based on the interrelationships between disciplines, noting that there may be many more.

In addition to the diversity of ways in which the term "STEM" itself is perceived, the nature of STEM education around the world is also very diverse. Like educational events in general, STEM education events are social and

DOI: 10.4324/9781003392545-4

historical events. In other words, the nature of STEM education can change and be defined by countries, groups, and organizations on the one hand and by the times on the other (cf. Ohtaka, 2012, p. 180). It is precisely because of this diverse nature of educational events that international comparative and historical research on STEM education can be significant.

Aim and method

From a comparative educational research perspective, this chapter focuses on Japan and Germany. Japan has a historical background of being influenced by Western countries such as Germany, the United Kingdom, and the United States during and after the modernization of education in the *Meiji* era. Japan and Germany are non-English-speaking countries located in the East and West, respectively, and both have implemented STEM education in distinctive ways. Both countries are part of the global trend toward competency-based education reform, which emphasizes the competencies that students should acquire as learning outcomes. Given these circumstances, what types of student competencies are cultivated in STEM education in Japan and Germany? This chapter focuses mainly on the student competencies that STEM education aims to cultivate, and attempts to identify some of their characteristics through a comparative analysis of Japan and Germany. It should be noted that the term STEM is not generally used in Germany (Tönnsen, 2022, p. 100), and in the German context, the term MINT will be used primarily in the following discussion.

STEM education in Japan and MINT education in Germany can be categorized in a variety of ways depending on the stage of life they target and even for those aimed at children and students, taking into account differences arising from whether they are positioned within or outside the framework of school education. Moreover, the perspectives of comparative analysis may vary depending on various factors that constrain and regulate the nature of these educational activities. In light of this, and because the limited space of this chapter does not allow for an exhaustive discussion of the various aspects of STEM/MINT education in Japan and Germany, three basic perspectives are adopted for comparative analysis. As mentioned earlier, this chapter focuses on the third perspective. The first and second perspectives are based on matters that could, to varying degrees, directly or indirectly influence the analysis and arguments of the third perspective.

The first perspective is the background to the demand for STEM/MINT education. As global interest in STEM education has expanded in recent years, in what social or educational contexts are the needs for STEM/MINT education being questioned in both Japan and Germany? In general, it is posited that the purpose of STEM education has two major aspects: first, to develop STEM human resources as a nation specializing in the fields of science and technology; and second, to cultivate STEM-literate citizens living in today's increasingly uncertain society (e.g., Isozaki & Isozaki, 2021, pp. 150–151;

Ministry of Education, Culture, Sports, Science and Technology [MEXT] (Central Council for Education), 2021, p. 56; Renn et al., 2012, p. 17). Which aspects of STEM/MINT education have gained more attention in Japan and Germany? To resolve this question, I conducted a comparative analysis using various policy documents and position papers from STEM/MINT-related initiatives as central clues.

The second perspective considers STEM/MINT subjects and the subject areas offered in school education. Schools are not the only places where STEM/MINT education is offered. However, this discussion has been limited to school education, particularly secondary education. The reason for this focus on the secondary education stage is that it includes the final stage of compulsory education and is a time when children are faced with a choice of career paths. Thus, at this educational stage, the role of STEM/MINT education is drawing increasing attention, both in terms of fostering human resources as professionals/vocational experts and STEM-literate citizens. When considering STEM/MINT education in schools, what subject or subject area frameworks are currently being implemented or are envisioned for future implementation? While the idea of STEM "integration" is still being discussed in various ways (Moore et al., 2020), are there standalone subjects and subject areas that implement STEM/MINT in an integrated and co-curricular manner? Do no such subjects or subject areas exist, and is implementation based on existing subjects? Keeping in mind that major differences exist in the underlying educational systems of Japan and Germany, I compare and analyze the frameworks of STEM/MINT subjects and subject areas using the curriculum standards for general education established at the national or state level (*Länder*) level as a starting point.

The third perspective concerns student competencies cultivated through STEM/MINT education. The competency-based curriculum reforms developed in Japan, Germany, and other countries tend to place more emphasis on output as a learning outcome. How are student competencies cultivated through STEM/MINT education embodied and presented as output? Some research also defines "STEM competence" as "an individual's ability to apply STEM knowledge, skills, and attitude appropriately in his or her everyday life, workplace or educational context" (Boon Ng, 2019, p. 11). On the other hand, as competency-based curricula are being advocated, it has been pointed out that although the STEM education framework is being expanded in terms of "STEM skills," which differ from expertise in the traditional sense, the definition of the term "STEM skills" is unclear (Tytler, 2020, p. 27). In any case, it is important to consider the kind of universal/cross-disciplinary characteristics emphasized in STEM/MINT education and how these relate to the competencies that students are expected to develop in existing subject areas, and so I comparatively analyze competencies that are expected to be cultivated through STEM/MINT education with reference also to the general education curriculum standards established at the national or state level.

Some aspects of STEM education in Japan

Background to the demand for STEM education

The rapid technological innovation represented by IoT and AI, as well as the unprecedented COVID-19 pandemic that the world has been facing, have made the need for social structural reform even clearer. In this era of the fourth Industrial Revolution, Japan has been calling for the realization of "Society 5.0" (Cabinet Office, 2016). Society 5.0, which was proposed in the 5th Science and Technology Basic Plan approved by the Cabinet in 2016, is the future society that Japan should aim to become. It refers to "a human-centered society that balances economic advancement with the resolution of social problems by a system that highly integrates cyberspace and physical space" (Cabinet Office, n.d.). During this period of social transformation, new approaches to education are now coming into question (Ohtani, 2021, p. 94).

STE(A)M education is gaining traction as one such new approach to education, and looking ahead to the realization of Society 5.0, its necessity is becoming more evident. For example, in "Human Resource Development for Society 5.0: Changes to Society, Changes to Learning" announced by the MEXT in June 2018, the need for all students at the high school/upper secondary school level to undertake STEAM education, which is foundational to thinking, was clearly stated, and the necessity of producing outstanding STEAM human resources, world-leading researchers, and human resources capable of creating new value was emphasized (MEXT, 2018b, p. 13). Furthermore, the Central Council for Education report released in January 2021 highlighted the importance of promoting and enhancing STEAM education, especially at the upper secondary school level, under the new Course of Study as National Curriculum Standard (MEXT (Central Council for Education), 2021, pp. 56–58). As this report considers the purpose of STEAM education from the twin aspects of human resource development and the development of modern citizens, it focuses not only on the human resource development aspect but also on the citizen development aspect of STEAM education, taking the view that it is important to define and promote the "A" of STEAM in a broader scope (as Liberal Arts; MEXT (Central Council for Education), 2021, pp. 56–57). In Japan, the National Curriculum Standards for elementary, junior high/lower secondary, and upper secondary schools were revised between 2017 and 2018. Although the terms STEM and STEAM do not appear in the National Curriculum Standards, it is evident that the new curricula aim to promote STEAM education.

On the other hand, the Ministry of Economy, Trade and Industry (METI), which is responsible for the economic and industrial sectors, has also been making efforts to promote STEAM education. One leading example is the "Future Classrooms" project, launched in 2018 to examine new educational approaches in line with the changing times. The "Future Classrooms" vision, released in 2019, recognizes that human resource development toward Society 5.0 has not kept pace, and it sets the "promoting STEAM in learning" as one of its cornerstone policies in order to nurture each child into a change-maker who will create the future (METI, 2019). This "promoting STEAM in learning" advocates for learning that cycles between the acquisition of special-ized knowledge, irrespective of whether in humanities or sciences, and the positing and resolution of as yet unknown problems through creative and logical thinking (METI, 2019, p. 2). The need to promote STEAM education in cooperation with industry is also pointed out in the aforementioned Central Council for Education report (MEXT (Central Council for Education), 2021, p. 58).

As outlined above, in Japan, STE(A)M education is needed to further the development of world-leading STEM human resources and the cultivation of literate citizens who will live in such a society, mainly toward the realization of the new society known as Society 5.0.

STEM subjects and subject areas at the secondary education level

In Japan, the Enforcement Regulations for the School Education Law stipu-late the subjects that make up the curriculum. What types of subjects and subject areas are currently available in STEM education? Of course, with the STEAM approach, including the arts, the range of subjects and subject areas will be considerably broader. However, in this chapter, I will limit my discus-sion to those corresponding to the S, T, E, and M disciplines to some extent.

The current lower secondary school curriculum consists of nine subjects, the special subject "morality period," period for integrated studies, and *tok-katsu* (student-led activities) (MEXT, 2018a, p. 11). These include no in-dividual subjects specifically designated as "STEM." STEM-related subjects include science, mathematics, and technology and home economics (par-ticularly in the field of technology). A subject area directly corresponding to "Engineering" has not been established, but content related to information technology and programming is covered in some of the fields of technology in technology and home economics (MEXT, 2018a, pp. 134–135). Although these subjects are indeed related to various STEM disciplines, learning activities in the sense of "integrated STEM" are not explicitly positioned within them. However, period for integrated studies mentioned above places its pri-mary focus on cross-disciplinary and integrated learning, in which STEAM education is also expected to be implemented (MEXT, 2022).

Upper secondary school curricula have three broad divisions with sub-ject areas broadly divided into "subject areas common to each division" and "subject areas offered mainly in specialized divisions" (MEXT, 2019a, p. 11). Within these subjects, there are subjects belonging to each subject area, and the upper secondary school curriculum comprises these subjects, period for inquiry-based cross-disciplinary study, and *tokkatsu* (MEXT, 2019a, p. 7). In this section, I focus on the subject areas common to each division and examine their structures. As with lower secondary schools, there are no independent "STEM" subjects or subject areas in upper secondary schools. STEM-related subjects and subject areas include science (science and human life, basic physics, advanced physics, basic chemistry, advanced chemistry, basic biology, advanced biology, basic earth science, advanced earth science), mathematics (mathematics I, II, III, A, B, C), informatics (informatics I, II), and inquiry-based study of science and mathematics (basic inquiry-based study of science and mathematics, inquiry-based study of science and mathematics). However, not all these are compulsory. Moreover, there are subjects related to "Technology" in the specialized divisions, but not in the general education divisions. Noteworthy among the subjects and subject areas listed above are basic inquiry-based study of science and mathematics and inquiry-based study of science and mathematics, which were newly established in the 2018 revision of the upper secondary school national curriculum standard. Although not mandatory, these are inquiry-based subjects that encompass mathematics and science, and can be described as aligned with the notion of STEM education (MEXT, 2019b, p. 7). In addition, period for inquiry-based cross-disciplinary study, which builds on period for integrated studies from lower secondary school and emphasizes more inquiry-based activities, is offered. Period for inquiry-based cross-disciplinary study and other subjects such as inquiry-based study of science and mathematics have a strong affinity with STEAM education in that they address real-world cross-curricular issues and integrate learning across individual subject areas for inquiry to solve problems, and they are expected to occupy a central position for STEAM education in upper secondary schools (MEXT (Central Council for Education), 2021, p. 57).

To briefly summarize, these are the key points about STEM subjects in Japan: (1) there are no independent subjects bearing the name "STEM;" (2) subjects corresponding to "S" and "M" are offered through lower secondary and upper secondary school curricula, but for "T" and "E," however, this is not always the case; (3) courses have been established for cross-disciplinary and integrated learning (inquiry-based study of science and mathematics, period for inquiry-based cross-disciplinary study, etc.), in which STEAM education is expected to be implemented.

Competencies to be cultivated in STEM education

Japan's National Curriculum Standards revised in 2017 and 2018 are strongly oriented toward "competency (in Japanese, *shishitsu-nouryoku*)-based" reform (Ishii, 2017). The three cornerstone competencies that children require in order to shape their future have been identified as "knowledge and skills," "abilities to think, make judgement and express themselves," and "motivation to learn, and humanity" (MEXT (Central Council for Education), 2016, pp. 26–31). Based on these three cornerstones, the goals and contents of each subject and subject area were restructured.

As mentioned earlier, in Japan, there are no independent subjects designated as "STEM," and it is not necessarily the case that the competencies to be cultivated through STEM education are always specified at the level of the National Curriculum Standard as part of the official curriculum. Therefore, although somewhat indirect, I will explore the competencies to be cultivated by examining the objectives of subjects such as inquiry-based study of science and mathematics and period for inquiry-based cross-disciplinary study, which are expected to be implemented as STE(A)M education. Table 3.1 summarizes the objectives of inquiry-based study of science and mathematics and period for inquiry-based cross-disciplinary study. Note that 1, 2, and 3 in the table describe the competencies corresponding to "knowledge and skills," "abilities to think, make judgement and express themselves," and "motivation to learn, and humanity," respectively. Table 3.1 shows that, as per the underlined sections, the central aim is to develop competencies to discover and solve problems related to various phenomena (including those in the contexts of society and real life). In developing these competencies, it is assumed that the process of inquiry will be conducted while comprehensively and integrally

Table 3.1 Objectives of the subjects "inquiry-based study of science and mathematics" and "period for inquiry-based cross-disciplinary study"

Inquiry-based study of science and mathematics	*Period for inquiry-based cross-disciplinary study*
Through the process of inquiry by relating to various events and working with a combination of mathematical and scientific discipline-based epistemological approaches, this subject aims to cultivate the following basic <u>competencies necessary to solve problems</u>:	Through the use of discipline-based epistemological approaches to inquiry, and through cross-disciplinary and integrated learning, this subject aims to develop the <u>competencies to better discover and solve problems</u>, taking into account one's own way of being and living, as follows:
1 Acquire the knowledge and skills necessary to inquire into events. 2 Perceive events from multiple and complex perspectives, set and inquire into problems related to mathematics, science, etc., and develop the ability to solve problems and enhance creative abilities. 3 Develop an attitude to face various events and problems proactively, to think and act with persistence, to actively take on the challenge of solving problems and create new values, and to evaluate and improve the process of inquiry in retrospect, as well as to develop an ethical attitude.	1 Through the process of inquiry, acquire the knowledge and skills necessary to discover and solve problems, form concepts related to problems, and understand the significance and value of inquiry. 2 Discover questions based on the relationship between oneself and the real world and real life, formulate problems independently, gather information, organize and analyze it, and summarize and express the results. 3 Develop an attitude to work proactively and collaboratively in inquiry, and to create new values and realize a better society while making the most of each other's strengths.

Note: Prepared by the author based on MEXT (2019a, pp. 197, 475; underlining by the author)

applying discipline-based epistemological approaches characteristic of each subject area. Moreover, it can be seen that these competencies will transcend the boundaries of the subjects themselves. In addition to these problem-solving skills, the MEXT (Central Council for Education) (2021, p. 57) argues the importance of taking advantage of the characteristics of STEAM education from a cross-disciplinary perspective to cultivate the competencies that form the foundation of learning, such as language- and information-utilizing skills, as well as the competencies required to respond to various contemporary problems.

As outlined above, the competencies that STEM education aims to cultivate are not always explicitly stated but can be viewed as general-purpose and cross-disciplinary competencies, such as the ability to identify and solve problems in the context of the real world.

Some aspects of MINT education in Germany

Background to the demand for MINT education

In Germany, which also advocates *Industrie 4.0* (the fourth Industrial Revolution), many MINT programs and initiatives have appeared in the past 20 years or so, and MINT education has taken root in a short period of time (Tönnsen, 2022, p. 119).

In 2019, the German Federal Ministry of Education and Research (*Bundesministerium für Bildung und Forschung*: BMBF) announced the MINT action plan "Creating the future with MINT!" to strengthen MINT education, stating the following: "If Germany wants to stay at the forefront of science, technology and innovation, we need more people who are interested in mathematics (*Mathematik*), computer science (*Informatik*), natural sciences (*Naturwissenschaften*) and technology (*Technik*) – MINT for short" (BMBF, 2019, p. 2). In other words, the need for MINT education is discussed in terms of Germany's leading position in global competition. Connected to this, a particularly significant problem that has been widely pointed out is the shortage of experts and labor in MINT fields in Germany (e.g., BMBF, 2012, pp. 26–27; BMBF, 2019, p. 5). In addition to imbalances between supply and demand, there is a lack of sustained interest in MINT among young people, who represent future professionals and the workforce in these fields, as well as a lack of women in MINT fields (BMBF, 2019, p. 5). As these issues are being addressed, MINT education is being promoted from the standpoint of developing MINT human resources, which can contribute to Germany's global competitiveness and social prosperity. However, MINT education is certainly not promoted solely with the human resource development aspect in mind. For example, the earlier MINT action plan by the BMBF established "MINT in Society" as one of the four action areas for strengthening MINT education, and states that a basic understanding of science and technology is widely required by the general public (BMBF, 2019, p. 22). Thus, it is evident that

MINT education is in demand to cultivate citizens with a basic understanding of MINT.

Incidentally, the trend of MINT education in Germany is not unrelated to the educational reforms in Germany since the 2000s following the so-called "PISA shock": the shocking results of Germany's participation in the OECD-PISA survey. Take, for example, the 2009 resolution of the Standing Conference of the Ministers of Education and Cultural Affairs of the Länder in the Federal Republic of Germany (*Kultusministerkonferenz*: KMK), regarding "*Recommendations to strengthen mathematics, natural sciences and technical education.*" In addition to asserting the value of MINT education, this resolution mentions measures to strengthen mathematics, natural sciences, and technology education. This includes a thorough pivot toward the educational standards (*Bildungsstandards*) developed by the KMK and the development of competency-oriented teaching in MINT subjects (Kultusministerkonferenz (KMK), 2009). Interestingly, the post-PISA reform of mathematics, science, and technology education is positioned within the trend of promoting MINT education. Furthermore, Kennedy-Salchow (2018) notes that the decline in Germany's traditional dual vocational training system and concerns about the future workforce led to industry (companies, their respective foundations, etc.) becoming actively involved in general education following the PISA shock, with their focus narrowing to MINT education. This may be one of the reasons why German MINT education is often characterized by an abundance of extracurricular learning opportunities provided by companies (Renn et al., 2012, pp. 20–21).

In any case, it can be said that MINT education in Germany aims both to develop MINT professionals as future experts in MINT fields and to foster citizens with a basic understanding of MINT.

MINT subjects at the secondary education level

As is widely known, in Germany the authority over education and culture is basically held by the states, not the federal government (KMK, 2021, p. 15). This has led to diversity in the establishment and naming of subjects by state, and even by school type. On the other hand, as mentioned earlier, educational reforms after the PISA shock led to the formulation of educational standards at the federal level by the KMK. These standards "address general educational objectives and define what competencies students should acquire by a certain grade level with respect to key content" (KMK, 2021, p. 381), but they were not formulated for all subjects. Regarding MINT subjects at secondary levels, the formulation of educational standards is limited to mathematics, physics, chemistry, and biology at both lower and upper secondary levels (see KMK, n.d.), whereas educational standards for technology and computer science have not been formulated. Of course, no educational standards exist for "MINT" as a standalone subject.

The situation with technology and computer science differs from state to state and from school to school. These are not necessarily established in all states (Tönnsen, 2022, p. 121). A survey by the German Mechanical Engineering Industry Association (*Verband Deutscher Maschinen- und Anlagenbau e.V.*; VDMA) reported that only a limited number of states include technology as an independent subject and that it is rarely taught in school types such as *Gymnasiums* (Verband Deutscher Maschinen- und Anlagenbau e.V. (VDMA), 2019, p. 27). Thus, even for the same MINT subjects, the situation is not uniform. Although a few subjects are mixed in nature, such as the subject "nature and technology" (see State Institute for School Quality and Educational Research, n.d.) in the first semester secondary stage in Bavaria (5th to 7th grades of the *Gymnasium*), for example, these are not specifically mentioned as the main site for developing integrated MINT education.

To summarize, the following are key points regarding MINT subjects in Germany: (1) there are no independent subjects designated as "MINT." (2) only mathematics, physics, chemistry, and biology are included in the federal educational standards established by the KMK, whereas technology and computer science are not. In addition, the status of technology and computer science as subjects varies from state to state.

Competencies to be cultivated in MINT education

Because MINT does not exist as a standalone school subject, it can be difficult to directly identify the competencies that MINT education aims to cultivate within the framework of school education. The MINT action plan published by the BMBF in 2019 also uses the term "MINT-competence," but does not clearly explain what it refers to. On the other hand, as mentioned earlier, if we consider the educational development in MINT-related subjects after the PISA shock as formative of MINT education, it is worth paying attention to the competencies indicated as achievement goals in the KMK educational standards.

Let us focus mainly on the educational standards for the *Allgemeine Hochschulreife* (general higher education entrance qualification) in physics, chemistry, and biology (corresponding to the upper secondary education stage), which were adopted by the KMK in 2020. These standards specify the competencies to be cultivated by teaching each subject. The concept of competence itself is described in the following article: "Competence is defined as the ability to apply knowledge and skills in the relevant subjects to solve problems" (KMK, 2020a, 2020b, 2020c, p. 3). First, it should be pointed out that the terms MINT or MINT-competence does not appear at all in any of the educational standards for physics, chemistry, or biology. The same can be said of the educational standards for the *Allgemeine Hochschulreife* in mathematics, which were adopted by KMK in 2012 (KMK, 2012). The education standards for physics, chemistry, and biology have four areas of competence (content knowledge, scientific reasoning, communication, and decision-making) in common with these three subjects (KMK, 2020a, 2020b, 2020c, p. 10), but the individual competencies to be developed are phrased in a way that strongly reflects the unique aspects of each subject. Such competencies include interdisciplinary relationships and cross-disciplinary perspectives: for example, "The students employ cross-disciplinary approaches to the interpretation of investigation findings" (KMK, 2020b, p. 16). However, these remain limited in scope.

Either way, in Germany, MINT education within the framework of school education has a strong M, I, N, and T approach centered on existing subjects, although the positioning of each subject differs, and the competencies to be cultivated are mainly defined by aspects that are specific to each subject. The status of general-purpose, cross-disciplinary competencies that transcend the framework of MINT subjects remains unclear.

Discussion

Thus far, I have explored the reality of STEM/MINT education in Japan and Germany, albeit from a few perspectives. A comparison of Japanese and German STEM/MINT education based on the three basic perspectives outlined above reveals several distinctive points.

First, looking at the background to the demand for STEM/MINT education and focusing on discussions at the educational policy level, clearly, both Japan and Germany are looking ahead to a rapidly changing society, as expressed in key phrases such as Society 5.0 and *Industrie 4.0*. I will also focus on two key aspects: the first is the national policy of cultivating STEM/MINT personnel who will lead this society, and the other is developing STEM/MINT literacy for all citizens living in this society. Of course, there are some differences in the emphasis placed on these two key aspects, but a certain commonality can be found between Japan and Germany in that the advancement of STEM/MINT education is sought from both perspectives. In relation to this, Isozaki and Isozaki (2021, p. 150) have argued that the "for all" perspective (i.e., improving the scientific and technological literacy of the wider population) should be combined with the "for excellence" perspective (i.e., the training of future experts in STEM/MINT fields) to comprise a hybrid curriculum. Given the demand for STEM/MINT education in both countries, the question of how to embody and implement such a curriculum in the context of STEM/MINT education will continue to be an ongoing challenge.

Regarding STEM/MINT subjects in school education (at the secondary school level), one commonality between Japan and Germany is that there are no independent subjects named STEM or MINT. This is consistent with the finding of van Driel et al. (2018, p. 33) that STEM is not a single subject or course worldwide. In addition, while mathematics and science are set as compulsory subjects, some subjects equivalent to technology, engineering, and informatics are not, and this application is not uniform. One characteristic of Japan is that its curricula include subjects for cross-disciplinary and comprehensive learning, such as inquiry-based study of science and mathematics and period for inquiry-based cross-disciplinary study, in which STEAM education is expected to be implemented. In other words, there is a framework for integrated STEM education in schools, although it is not specifically called STEM. By contrast, in Germany, integrated MINT education is mostly implemented outside existing academic classes (Tönnsen, 2022, p. 121). Vasquez et al. (2013, p. 72) provide a typology of approaches to STEM curriculum design based on the degree of integration: disciplinary, multidisciplinary, interdisciplinary, transdisciplinary. It is not difficult to imagine that these differences in approaches must also affect the sites in which they are implemented, which, in the case of school education, are primarily subjects or subject areas, and how they are implemented. We must consider whether this approach can be implemented in existing subjects and subject areas, whether it requires a new subject framework, and whether it is better to seek a place outside school education to implement the approach. We may need to question how subjects and subject areas are organized according to approach. In addition, it is important to consider the implementation of STEM education from the perspective of "collaboration" among the various subjects.

Although there is not necessarily a clear definition of the competencies to be cultivated in STEM/MINT education in either Japan or Germany, in Japan, the focus seems to be more on versatility and cross-disciplinarity (e.g., the ability to identify and resolve problems in the real world), whereas in Germany, the focus seems to be more on aspects specific to each subject. These differences are related, in no small measure, to the differences in STEM/MINT approaches in terms of the degree of integration (compartmentalized vs. integrated). Of course, the question is not which of the above approaches is better or worse for competency cultivation in STEM/MINT education. However, based on a comprehensive review of the discussion thus far, there are further points to consider when examining the competencies students should acquire through STEM education. The key points are not limited to those listed below; however, it is worth noting two main points.

The first is to reexamine the relationship between cross-disciplinary competencies and the competencies specified in each STEM subject base. Generally, when responding to real-world problems from a cross-disciplinary perspective, it is necessary to utilize individual and specialized knowledge and skills acquired through the study of individual subjects and mobilize them in response to the context. Rather than viewing cross-disciplinary competencies as a mere "collection" of STEM subject-based competencies, it may be significant to reconsider the nature of cross-disciplinary competencies by focusing on the meta-aspect of "mobilizing" these competencies appropriately, according to the context of the task at hand. Furthermore, in the process of "mobilizing," students are exposed to differences in the distinctive views and approaches of several subjects, which can lead to a deeper understanding of the distinctive aspects and limitations of each subject's views and approaches. Although one characteristic of integrated STEM is that it addresses complex real-world issues (Mohr-Schroeder et al., 2020, p. 30), the significance of STEM education can also be found in aspects other than the cultivation of problem-solving skills.

The second is to question the subject framework for cultivating competencies, which is also related to the aforementioned point. The current lack of a standalone STEM subject worldwide (van Driel et al., 2018, p. 33) indicates the difficulty of establishing such a subject. However, when facing the issue of the aims and objectives of STEM education, such as the competencies that students should acquire through STEM education, the question of the place (subject framework) in which these competencies should be cultivated is also closely implicated. For example, can cross-disciplinary competencies be cultivated in existing STEM subjects? Is it appropriate to completely segregate cross-disciplinary competencies into subject areas that specialize in cross-disciplinary or integrated learning, while competencies defined in individual STEM subject areas are developed in existing STEM subject areas? While the division and integration of subjects is not a binary choice, such questions, which also relate to the foundations on which these subjects are based, must be considered.

Conclusion

As noted in the Introduction, the fact that STEM education has become a global trend in recent years cannot be disputed. However, the reality of STEM education in different countries and cultures is not necessarily uniform, as is the case in Japan and Germany (e.g., Marginson, 2015). Given this context, the question is how to interpret the concept of STEM in the context of one's own country and create specific forms of education. The findings of comparative education research, such as this, will certainly be crucial to such attempts.

This chapter has focused on the aims and objectives of STEM education, particularly the question of "What are the student competencies to be cultivated in STEM education?," basing its discussion on comparisons between Japan and Germany. These competencies can include a wide range of aspects, from those that focus on aspects specific to each subject to those that focus on cross-disciplinary aspects. It is also important to focus on the content and methods of STEM education, such as "what" is taught in STEM education and "how" it is taught. However, given that the aims and objectives of (STEM) education define the content and methods to some extent (Ohtaka, 2012, pp. 181–183), when comparing STEM education content and methods, it is important to be at least conscious of the aspects of the aims and objectives of STEM education.

Acknowledgments

This work was supported by the Japan Society for the Promotion of Science KAKENHI Grant Number JP21H00919 (JP23K20744).

References

Boon Ng, S. (2019). *Exploring STEM competences for the 21st century.* https://unesdoc.unesco.org/ark:/48223/pf0000368485

Bundesministerium für Bildung und Forschung (BMBF). (2012). *Perspektive MINT. Wegweiser für MINT-Förderung und Karrieren in Mathematik, Informatik, Naturwissenschaften und Technik [MINT outlook. Guidebook for promoting MINT and careers in mathematics, computer science, natural sciences and technology].* https://www.yumpu.com/de/document/read/493882/perspektive-mint [in German].

Bundesministerium für Bildung und Forschung (BMBF). (2019). *Mit MINT in die Zukunft! Der MINT-Aktionplan des BMBF [Creating the future with MINT! The MINT action plan created by the Federal Ministry of Education and Research].* https://www.bmbf.de/SharedDocs/Publikationen/de/bmbf/1/31481_Mit_MINT_in_die_Zukunft.pdf?__blob=publicationFile&v=8 [in German].

Bybee, R. W. (2013). *The case for STEM education: Challenges and opportunities.* NSTA Press.

Cabinet Office. (n.d.). *Society 5.0.* Retrieved May 2, 2023, from https://www8.cao.go.jp/cstp/english/society5_0/index.html

Cabinet Office. (2016). *Dai goki kagaku-gijutsu-shinkō-keikaku [The 5th science and technology basic plan].* https://www8.cao.go.jp/cstp/kihonkeikaku/5honbun.pdf [in Japanese].

Ishii, T. (2017). *Chūkyōshin "tōshin" wo yomitoku: Shin gakusyū shidō yōryō wo tsukai-konashi, shitsu no takai jugyō wo sōzō suru tameni [Reading and understanding the report of the Central Council for Education: Mastering the new National Curriculum Standard and creating high-quality classes].* Nipponhyōjun [in Japanese].

Isozaki, T., & Isozaki, T. (2021). Nihongata STEM-kyōiku kōchiku ni mukete no rironteki kenkyū: Hikaku kyōikugakuteki shiza kara no bunseki wo tōshite [Theoretical research for establishing a Japanese-style STEM education: Analysis from a comparative historical point of view]. *Journal of Science Education in Japan, 45*(2), 142–154. https://doi.org/10.14935/jssej.45.142 [in Japanese with English summary].

Kennedy-Salchow, S. (2018). *Corporate philanthropy practices in K-12 education in the U.S. and Germany* (Order No. 27732656). Available from ProQuest Dissertations & Theses Global. (2425008988). https://www.proquest.com/dissertations-theses/corporate-philanthropy-practices-k-12-education-u/docview/2425008988/se-2

Khine, M. S., & Areepattamannil, S. (Eds.). (2019). *STEAM education: Theory and practice.* Springer.

Kultusministerkonferenz (KMK). (2009). *Empfehlung der Kultusministerkonferenz zur Stärkung der mathematisch-naturwissenschaftlich-technischen Bildung (Beschluss der Kultusministerkonferenz vom 07.05.2009) [Recommendations to strengthen mathematics, natural sciences and technical education (resolution reached by the KMK on May 7, 2009)].* https://www.kmk.org/fileadmin/Dateien/veroeffentlichungen_beschluesse/2009/2009_05_07-Empf-MINT.pdf [in German].

Kultusministerkonferenz (KMK). (2012). *Bildungsstandards im Fach Mathematik für die Allgemeine Hochschulreife (Beschluss der Kultusministerkonferenz vom 18.10.2012) [Educational standards for the subject mathematics at the general tertiary educational level (resolution reached by the KMK on October 18, 2012)].* https://www.kmk.org/fileadmin/Dateien/veroeffentlichungen_beschluesse/2012/2012_10_18-Bildungsstandards-Mathe-Abi.pdf [in German].

Kultusministerkonferenz (KMK). (2020a). *Bildungsstandards im Fach Biologie für die Allgemeine Hochschulreife (Beschluss der Kultusministerkonferenz vom 18.06.2020) [Educational standards for the subject biology at the general tertiary educational level (resolution reached by the KMK on June 18, 2020)]*. Wolters Kluwer [in German].

Kultusministerkonferenz (KMK). (2020b). *Bildungsstandards im Fach Chemie für die Allgemeine Hochschulreife (Beschluss der Kultusministerkonferenz vom 18.06.2020) [Educational standards for the subject chemistry at the general tertiary educational level (resolution reached by the KMK on June 18, 2020)]*. Wolters Kluwer [in German].

Kultusministerkonferenz (KMK). (2020c). *Bildungsstandards im Fach Physik für die Allgemeine Hochschulreife (Beschluss der Kultusministerkonferenz vom 18.06.2020) [Educational standards for the subject physics at the general tertiary educational level (resolution reached by the KMK on June 18, 2020)]*. Wolters Kluwer [in German].

Kultusministerkonferenz (KMK). (2021). *The education system in the Federal Republic of Germany 2018/2019. A description of the responsibilities, structures and developments in education policy for the exchange of information in Europe.* https://www.kmk.org/fileadmin/Dateien/pdf/Eurydice/Bildungswesen-engl-pdfs/dossier_en_ebook.pdf

Kultusministerkonferenz (KMK). (n.d.). *Bildungsstandards der Kultusministerkonferenz [KMK educational standards]*. Retrieved May 28, 2023, from https://www.kmk.org/de/themen/qualitaetssicherung-in-schulen/bildungsstandards.html#c5034 [in German].

Marginson, S. (2015). What international comparisons can tell us. In B. Freeman, S. Marginson, & R. Tytler (Eds.), *The age of STEM: Educational policy and practice across the world in science, technology, engineering and mathematics* (pp. 22–32). Routledge.

Ministry of Economy, Trade and Industry (METI). (2019). *"Mirai no kyōshitsu" bijon: Keizaisangyōsyō "mirai no kyōshitsu" to EdTech kenkyūkai dai niji teigen ["Future Classrooms" Vision: Ministry of Economy, Trade and Industry's "Future Classrooms" and EdTech Study Group's Second Proposal]*. https://www.meti.go.jp/shingikai/mono_info_service/mirai_kyoshitsu/pdf/20190625_report.pdf [in Japanese].

Ministry of Education, Culture, Sports, Science and Technology (MEXT). (2018a). *Chūgakkō gakusyū shidō yōryō (heisei 29 nen kokuji) [The Course of Study as National Curriculum Standard for lower secondary school (announced in 2017)]*. Higashiyama Shobō [in Japanese].

Ministry of Education, Culture, Sports, Science and Technology (MEXT). (2018b). *Society 5.0 ni muketa jinzaiikusei~Shakai ga kawaru, manabi ga kawaru~ (Human resource development for Society 5.0: Changes to society, changes to learning)* [in Japanese]. Retrieved from https://www.mext.go.jp/component/a_menu/other/detail/__icsFiles/afieldfile/2018/06/06/1405844_002.pdf

Ministry of Education, Culture, Sports, Science and Technology (MEXT). (2019a). *Kōtōgakkō gakusyū shidō yōryō (heisei 30 nen kokuji) [The Course of Study as National Curriculum Standard for upper secondary school (announced in 2018)]*. Higashiyama Shobō [in Japanese].

Ministry of Education, Culture, Sports, Science and Technology (MEXT). (2019b). *Kōtōgakkō gakusyū shidō yōryō (heisei 30 nen kokuji) kaisetsu Risūhen [Guideline for the Course of Study as National Curriculum Standard for Upper Secondary School (announced in 2018), Inquiry-Based Study of Science and Mathematics]*. Tokyo Shoseki [in Japanese].

Ministry of Education, Culture, Sports, Science and Technology (MEXT). (2022). *Ima, motomerareru chikara wo takameru sōgōteki na gakusyū no jikan no tenkai chūgakkō hen [Developing the Period for Integrated Studies to enhance the skills required now, lower secondary school version].* https://www.mext.go.jp/a_menu/shotou/sougou/20220426-mxt_kouhou02-2.pdf [in Japanese].

Ministry of Education, Culture, Sports, Science and Technology (MEXT) (Central Council for Education). (2016). *Yōchien, shōgakkō, chūgakkō, kōtōgakkō oyobi tokubetsushiengakkō no gakusyū shidō yōryō no kaizen oyobi hitsuyō na hōsaku tō ni tsuite (tōshin) [Improving the National Curriculum Standards and essential measures for kindergartens, elementary schools, lower secondary schools, upper secondary schools, and special needs schools (report)].* https://www.mext.go.jp/b_menu/shingi/chukyo/chukyo0/toushin/__icsFiles/afieldfile/2017/01/10/1380902_0.pdf [in Japanese].

Ministry of Education, Culture, Sports, Science and Technology (MEXT) (Central Council for Education). (2021). *"Reiwa no nihongata-gakkōkyōiku" no kouchiku wo mezashite: Subete no kodomotachi no kanousei wo hikidasu, kobetusaitekinamnabi to kyōdoutekina manabi no jitugen (tōshin) [Toward the construction of "Japanese style school education": Optimal individualized learning and collaborative learning that bring out the potential all children (report)].* https://www.mext.go.jp/content/20210126-mxt_syoto02-000012321_2-4.pdf [in Japanese].

Mohr-Schroeder, M. J., Bush, S. B., Maiorca, C., & Nickels, M. (2020). Moving toward an equity-based approach for STEM literacy. In C. C. Johnson, M. J. Mohr-Schroeder, T. J. Moore, & L. D. English (Eds.), *Handbook of research on STEM education* (pp. 29–38). Routledge.

Moore, T. J., Johnston, A. C., & Glancy, A. W. (2020). STEM integration: A synthesis of conceptual frameworks and definitions. In C. C. Johnson, M. J. Mohr-Schroeder, T. J. Moore, & L. D. English (Eds.), *Handbook of research on STEM education* (pp. 3–16). Routledge.

Ohtaka, I. (2012). Rika-kyōiku no mokuteki-mokuhyō to kagakuteki riterashī [Aims and objectives of science education and scientific literacy]. In I. Ohtaka & Y. Shimizu (Eds.), *Kyōka-kyōiku no riron to jugyō II risūhen [Theory of subject-based education and lessons II: Science and Mathematics]* (pp. 179–192). Kyōdōshuppan [in Japanese].

Ohtani, T. (2021). STEM/STEAM-kyōiku wo dou kangaereba yoika: Syogaikoku no dōkō to nihon no genjō wo tōshite [How to think about STEM/STEAM education: Trough trends in other countries and the current situation in Japan]. *Journal of Science Education in Japan, 45*(2), 93–102. https://doi.org/10.14935/jssej.45.93 [in Japanese with English summary].

Renn, O., Duddeck, H., Menzel, R., Holtfrerich, C. L., Lucas K., Fischer W., Allmendinger J., Klocke F., & Pfenning U. (2012). *Stellungnahme und Empfehlungen zur MINT-Bildung in Deutschland auf der Basis einer europäischen Vergleichsstudie [Opinions and recommendations for MINT education in Germany, based on a comparative European study].* Berlin-Brandenburgische Akademie der Wissenschaften [in German].

Sanders, M. (2009). STEM, STEM education, STEM mania. *Technology Teacher, 68*(4), 20–26.

State Institute for School Quality and Educational Research. (n.d.). *Natur und Technik [Nature and technology].* Retrieved May 28, 2023, from https://www.lehrplanplus.bayern.de/fachprofil/gymnasium/nt_gym/5 [in German].

Tönnsen, K.-C. (2022). Status and trends of STEM education in Germany. In Y. -F. Lee & L. -S. Lee (Eds.), *Status and trends of STEM education in highly competitive countries: Country reports and international comparison* (pp. 97–140). https://files. eric.ed.gov/fulltext/ED623352.pdf

Tytler, R. (2020). STEM education for the twenty-first century. In J. Anderson & Y. Li (Eds.), *Integrated approaches to STEM education: An international perspective* (pp. 21–43). Springer.

van Driel, J. H., Vossen, T. E., Henze, I., & de Vries, M. J. (2018). Delivering STEM education through school-industry partnerships: A focus on research and design. In T. Barkatsas, N. Carr, & G. Cooper (Eds.), *STEM education: An emerging field of inquiry* (pp. 31–44). Koninklijke Brill NV.

Vasquez, J. A., Sneider, C., & Comer, M. (2013). *STEM lesson essentials, grades 3–8: Integrating science, technology, engineering, and mathematics.* Heinemann.

Verband Deutscher Maschinen- und Anlagenbau e.V. (VDMA). (2019). *Technikunterricht in Deutschland. Eine Analyse und Bewertung von Technik in den Curricula allgemeinbildender Schulen [Technical education in Germany. An analysis and assessment of technology in the curricula for general schooling].* https://www.vdma.org/documents/34570/0/ VDMA+Kompendium+-+Technikunterricht+in+Deutschland_1567435852469.pdf/ f54af245-345d-4d7a-4ae9-cad4c1d38fdd?t=1612885653616 [in German].

4 Examination of STEAM teaching strategies in Japanese elementary education through a comparative analysis with STEAM learning activities in England

Susumu Nozoe

Introduction and aims

Western countries that support STEAM (science, technology, engineering, the arts, and mathematics) education, exhibit different types of learning activities based on their respective historical backgrounds and social circumstances. In England, for example, STEAM is not an official subject in the national curriculum, and only a few textbooks used for each subject address STEAM themes and topics. The STEAM educational system is similar to the one that is currently applied in Japan.

STEAM education in Japan aims to equip learners with the qualities and abilities necessary to become citizens who effectively contribute to modern society—a society deeply engrossed with STEAM concerns—and currently encourages cross-curricular learning to expand and deepen acquired knowledge, especially in high schools (Ministry of Education, Culture, Sports, Science and Technology [MEXT], 2021). Although the central academic and economic goals of science education are to encourage students to major in disciplines in the field of science by enhancing their interest in STEM and their motivation to continue studying it (Fortus et al., 2022), cross-curricular STEAM education at the high school level can be seen as an educational strategy that runs counter to the Japanese science curriculum, in which subject content becomes more specialized as the level of education increases.

The objective of this chapter is to clarify the characteristics and application of STEAM education in Japanese elementary schools through a comparative analysis with England and to examine the STEAM teaching strategies used in elementary education in Japanese schools. Hence, the following research questions have been formulated:

1 Do Japanese elementary schools' science textbooks contain activities related to STEAM education?
2 If so, what are the differences between the STEAM pedagogies present in Japanese elementary science textbooks, and the comparable textbooks used in other countries (England in this chapter)?

DOI: 10.4324/9781003392545-5

3 How should the STEAM educational approach be smoothly incorporated into elementary education in Japan? In other words, what teaching strategies are needed to successfully introduce it in Japanese elementary schools?

Why do we need STEAM education in elementary schools?

In this chapter, I begin by discussing why I specifically focused on elementary schools. STEAM education should be developed for a wide range of grades, from elementary to secondary education. However, as students proceed to secondary school, they are also influenced by other educational elements and considerations pertaining to their employment choices and options in the future. In particular, when considering the development of STEAM education and teaching strategies in formal education, such as at the secondary school level, it is necessary to discuss whether the target should be students who pursue advanced levels of education or employment in STEM fields (e.g., for excellence), or whether it should also include students who pursue higher education or employment preferences in non-STEM fields (that is, for all). However, STEAM education at the elementary level, even in formal education settings such as elementary schools, could in principle be based on a teaching strategy that assumes and incorporates all students (for all), in contrast to the case of secondary schools where specialization starts to emerge.

One of STEAM education's frequently expressed goals is to focus on developing competencies to solve a variety of increasingly complex challenges (e.g., alternative energy, sustainable resources, food shortages, and the detrimental effects of climate change). In the future, the current young generation will be expected to collaborate with others to create a sustainable and better society to the benefit of all in the harsh and challenging context of conflicting interests and values. The important perspective here is that such an improved future society will not be created solely by experts who have embarked on studies and careers in science and technology, but that the final democratic decision will be left to citizens who can understand the broad social and economic implications of those experts' points of view and arguments. It would not be an exaggeration to say that the key to a mature future society lies in the depth and number of the citizenry who have acquired these skills (Isozaki & Nozoe, 2017; Nozoe, 2015, 2019).

Research methods

Reasons for setting a comparative country

When performing a comparative analysis, one needs to analyze and compare based on the apposite dimensions and levels in order to arrive at meaningful results. After examining the most relevant aspects related to the current research subject, I decided to compare the educational systems of Japan with England for the following reasons:

1 Both Japan and England have unified and statutorily enacted curricula (i.e., the Course of Study in Japan, and the National Curriculum in England).
2 The textbooks in both countries are edited, based on the aforementioned statutory curricula.
3 In the Japanese elementary education system, science is taught in grades 3–6. England applies comparable school stages for those four years, also referred to as the 3rd–6th grades (i.e., Key Stage 2).

In order to make it easier to distinguish the notation of each country, "Primary" is used when referring to England. In addition, the grades are also indicated as "grade 3–6" for Japan and "3rd–6th grade" for England.

Study design

In the quest to answer my research questions, I first conducted a comparative analysis of elementary school science textbooks from Japanese five major companies (e.g., Arima et al., 2020a, 2020b, 2020c, 2020d; Ishiura et al., 2020a, 2020b, 2020c, 2020d; Mouri et al., 2020a, 2020b, 2020c, 2020d; Shimoda et al., 2020a, 2020b, 2020c, 2020d; Yoro et al., 2020a, 2020b, 2020c, 2020d), and science textbooks from England–Foxton Books (Tyrrell, 2020a, 2020b, 2020c, 2020d, 2020e, 2020f, 2020g, 2020h, 2020i, 2020j, 2020k, 2020l, 2020m, 2020n, 2020o, 2020p, 2020q, 2020r, 2020s, 2020t), all of which actively incorporate STEAM-inspired tasks. The analysis was conducted from two perspectives: (1) a quantitative perspective, such as which areas and grades are covered by STEAM activities and (2) a qualitative perspective, such as what specific content is covered by STEAM activities. By organizing and investigating these science textbooks in detail, the characteristics of STEAM activities in the elementary science textbooks of each country were empirically identified. Additionally, I discuss STEAM teaching strategies for elementary education in the context of proximal fields of research, with reference to the existing relevant literature.

It should be noted that in this survey, the focus is on "STEAM activities" rather than the vague and broad "STEAM education" in general (e.g., content and materials) so as to avoid ambiguity and any confusion in interpretation. STEAM activities are not specific to scientific inquiry such as observation and experimentation, but focus on activities with elements from other subjects and disciplines (STEAM elements). Moreover, given that the textbooks in question from both countries were published in the same year (2020), I did not factor their temporal backgrounds into the analysis.

Findings from the comparative analysis

Analysis of Japanese elementary school science textbooks

A comparative textbook analysis confirmed the activities that incorporated STEAM principles in all grades of Japanese elementary school science textbooks. However, the number of activities varied by grade, with most activities available at the grade 3 level. The following units involved activities in which students are required to make toys and tools, using scientific laws and properties: (i) the units of "the power of wind and rubber," "light," "sound," "electricity," and "magnet" in grade 3; (ii) the units of "the property of water and air," "current," and "temperature and volume of substances" in grade 4; (iii) the units of "pendulum" and "electromagnet" in grade 5; and (iv) the unit "leverage" in grade 6. This is a traditional and general method of education in Japanese elementary schools called, "manufacturing (*monozukuri*)," which had already been practiced in Japanese science education before the global STEAM movement. Unlike "general manufacturing," in which engineers and others express ideas as concrete objects or complete them as manufactured products, Japanese "manufacturing (*monozukuri*) in the subject of science" is a production activity positioned within the science lesson. It is characterized by the fact that it places great emphasis on the learning process, as well as the manufactured product as the outcome of the learning process. Examples of specific activities for each unit are as follows:

Grade 3

In "the power of wind and rubber" unit, activities were described for making objects such as cars and windmills that move with the force of wind and rubber, done from the perspective of converting the force of wind and rubber into motive power. In the "light" unit, a device that uses a plane mirror to make things brighter and warmer was made, and in the "sound" unit, an activity to produce a string telephone that transmits voices to a remote place, was completed and confirmed as such. In the "electricity" unit, activities such as making a switch with the aim of turning a miniature light bulb on and off, and manufacturing a tester with the aim of finding out if it conducts electricity, were described. In addition, the unit on "magnets" adopted activities such as making model cars and ships, using the properties of opposite and same poles of magnets.

Grade 4

In the unit on "the properties of water and air," activities to make air and water guns were described as examples, due consideration having been given to the fact that air has the property of being able to be compressed, whereas water cannot. In the "current" unit, activities to make a model car, a rotating swing, and a crane, whose movement changes depending on the number of batteries and the direction of the electric current, were also mentioned. Furthermore, in the "temperature and volume of substances," activities were identified in which thermometers were made by using the property of liquids to change in volume according to their temperature.

Grade 5

In the "pendulum" unit, the activity is described to create a simple metronome based on the pendulum principle. Also, in the unit on "electromagnet," activities were adopted to create an iron can pick-up machine and a coil motor, exploiting the property of becoming a magnet only when an electric current is flowing.

Grade 6

In the "leverage" unit, activities were described in which children made an object using the property of balance (e.g., a mobile sculpture made of wire, straw, string, etc. that, for instance, moves in the wind), and a pole scale that used the property of leverage. Additionally, in the unit "the utilization of electricity (power generation)," programming experiences that efficiently used electricity for the purpose at hand, were adopted. This reflects the programming education introduced in Japanese elementary schools in 2020. Through programming experience, logical ways of thinking (programming thinking) were advanced among the students, which skill they could then apply to programmed computers to perform intended processes.

Primary school science textbooks analysis for England

My analysis of Foxton's science textbooks in England revealed that STEAM activities were incorporated into all the grades under study. Unlike in Japan, these textbooks offer a similar number of activities for each grade; notably, they provide more than 20 activities for each grade. Such activities incorporate a wide range of STEAM elements, such as art and design, as well as technology and engineering. Many of the "STEAM activities" described in the science textbooks in England were activities that were aligned with each unit and mainly focused on "scientific experience," which is different from the way in which experiments were conducted in Japan. Some of the "STEAM activities" in the textbooks also included activities that would be carried out in Japanese subjects, such as arts and crafts, or home economics. Examples of specific activities for each unit are as follows:

3rd grade

In the "magnets and friction" unit, an activity was described in which children were free to draw pictures based on their understanding of the relationship between magnets and metals. This activity combines the science element of learning about magnets with the art element of drawing and is a unique activity in that it provides a fun way for children to learn.

4th grade

The "predators and prey" unit involved activities such as stacking paper cups with animal pictures to understand the structure of the food chain. Through this activity, the flow and structure of the food chain and organisms (producers and consumers) can be understood visually and experientially. In addition to the science component, this activity combines elements such as drawing and paper cup cutting, confirming the inclusion of technological and engineering components in STEAM.

5th grade

In the "earth and space" unit, an activity was described in which the children actually built a model of the solar system and tried to line it up, taking planets' distances from the sun into account. First, the children colored the models while looking at actual pictures of the solar system. At this junction of the exercise, the sun is 12 cm in diameter, Jupiter is 8 cm in diameter, and the larger planets have large spheres, whereas smaller planets, such as Mercury, have small spheres. Furthermore, the four inner planets were placed approximately 5 cm away from the sun, and the outer planets were approximately 8 cm apart, so that the children could sensibly determine the size of the planets and the distances between them. As it is not possible to see actual planets and bodies directly with their own eyes, model-making is the only learning activity that allows children to visualize the appearance of the solar system. Therefore, in subsequent activities, such as discussing what they notice when they look at the overall image of the solar system, or asking them to think about why the planets of the solar system do not move in a straight line, they are able to develop new questions and various ideas.

6th grade

In the "evolution and inheritance" unit, students can create original animals and ponder on how they have evolved while adapting to their environments. Prior to this, children are taught "what evolution is," "the theory of natural selection," and "types of adaptation." Therefore, they have some knowledge of how animals have evolved, and in this context, they build on this knowledge to create their own original animals. Among the requirements for this activity are "habitat resources such as websites and books." Using these resources, children investigate the kind of habitats animals lived in and how they have evolved. In the course of this activity, they consider which parts of the existing animals their original animals were made of, and how they had evolved. At the end of the learning activity, the children made posters of their animals, and shared them with the entire class. At that time, they presented detailed particulars about unique parts of the animals they had created.

Discussion

In summary, the above results clearly establish that STEAM activities are mainly adopted in the physics fields in Japanese elementary science education, and that they are not equally adopted across all fields, namely the disciplines of physics, chemistry, biology, and earth science. Very few manufacturing (*monozukuri*) activities were confirmed in units in the fields of biology and earth sciences, and they were not balanced across all fields. By contrast, England's primary science education has evenly adopted STEAM activities across all fields. While STEAM education in primary science in England focuses on learning activities in which students are invited to practice their creativity using art and design approaches that extend beyond the scope of science, Japan focuses on activities that connect children's senses to the formation of scientific concepts and activities that utilize scientific properties in elementary science education. Based on international trends, STEAM education in Japanese elementary schools may require new solutions and ideas based on free and interdisciplinary thinking. It should be noted, however, that not all the activities described in textbooks are actually carried out in elementary schools, so it is important to be careful not to assume that textbook content represents classroom practices in each country. The definition and goals of STEAM education vary from country to country; "innovation" often being used as a slogan to create new solutions and ideas. However, "innovation for scientists and engineers" and "innovation required by society and the world" are not exactly the same thing, and do not carry the same meaning. The former is evaluated by a group of experts (academic value) and does not necessarily serve as a driving force for economic growth expected by society and the world. Similarly, the latter considers science and technology for their ability to enrich world markets and society (economic and industrial value), and is not necessarily interested in inventions or discoveries of scientists and engineers in a relevant field.

At present, many educational researchers hold critical views on STEM education, arguing that it has a strong ideological component (e.g., Akerson et al., 2018; Takeuchi et al., 2020). Although the increasing trend of STEAM education provides evidence that educational systems are accepting the focus on human resource development in science and technology put forward by the economy and industry, this norm should not simply be accepted. Rather, it is important to incorporate STEAM into school education through practical knowledge, which Japan has long cultivated through its school-based pedagogy. As mentioned above, the Japanese "manufacturing (*monozukuri*) in the subject of science" is undergoing a review of its traditional features and trends, and is moving toward improvement. Rather than starting STEAM education by renewing the educational accumulation that has been built up by our predecessors, I would like to propose the transformation of "manufacturing (*monozukuri*) in the subject of science" into "STEAM activities" as a realistic teaching strategy in Japan. Many of the Japanese "manufacturing (*monozukuri*) in the subject of science" are positioned as developmental and applied learning at the end of a unit of study. The idea behind this approach is to deepen learners' understanding by producing concrete objects such as examples of the application of scientific principles and laws. For this reason, many elementary schools use teaching-resource kits for making things. In the case of making things with teaching material kits, all pupils make the same thing. From the teachers' point of view, it is easy to instruct, almost all pupils are expected to complete the work, and there is a certain degree of stability. However, I believe that there is little room for ingenuity and improvement in manufacturing, where everyone produces the same thing, and there are few opportunities for students to express new ideas and creativity. For example, as in the case study of England's textbooks, by giving some degree of freedom to the children by allowing them to automatically select materials, rather than using a kit, there is much more room for creativity from the conception and design stage of the object to be made, and it can be brought closer to a STEAM activity rather than just a "manufacturing experience." In addition, by setting an objective, such as solving some kind of problem, and introducing the perspective of trial and control of the operation of a manufactured product, it will be possible to review the characteristics and tendencies of conventional manufacturing. In this case, if the task is a narrative based on the context of the real-world and real-life, the activity may be more in line with STEAM's educational philosophy. It may also be a global trend in STEAM education, as programming education has been introduced into elementary education in Japan; however, technology and engineering in STEAM appear to be morphing into the development of computing skills. Children learn ICT faster than teachers, and there may be many situations in which teachers learn from learners. However, teachers must have the qualities and abilities (target axes) that their pupils want to develop. In their review of previous STEM-related articles, Nesmith and Cooper (2020) emphasized the importance of primary school teachers and curriculum developers considering appropriate STEM concepts and skills, and developing age-appropriate integrated STEM approaches as implications for teaching and curriculum design that must be considered when integrating STEM content in the actual primary classroom.

Specifically, this means incorporating the concept of engineering design, which is a different aspect from the traditional scientific inquiry, into the current "manufacturing (*monozukuri*) in science." Because other chapters on engineering design have been written, I will not go into detail and will only briefly describe the main points. Unlike scientific inquiry, which aims to elucidate natural phenomena, engineering design aims to solve problems in everyday life and the real world in front of us. In addition, although the answers obtained through scientific inquiry converge to a single conclusion, there are multiple solutions based on the engineering design process, and the cycle of redesign is repeated for new and improved solutions (e.g., Jolly, 2017). It is relatively easy for elementary school teachers, who teach all subjects, to incorporate these perspectives and elements into "manufacturing (*monozukuri*) activities in science," and the development of STEAM education in the context of Japanese elementary education is expected. The development of contextualized problem-solving learning, as science education in Japan has tended to focus on links between the real world and real life, will also support manufacturing (*monozukuri*) as a STEAM activity. For authentic STEAM education, it is important to relate to the real world, and it will be necessary to reconsider approaches such as *Socioscientific issues* and *Socioscientific reasoning* (Zeidler, 2016). Additionally, Wong et al. (2016) cautioned that the main focus of the current STEAM education in England is to increase the number of students who choose physics and mathematics subjects, and that the majority of students who do not choose a science-related course are ignored. Furthermore, they argued that if STEAM was to be positioned as an educational policy, reforms based on the perspective of *science education for all* would be necessary. By changing the problems to be solved to issues in the region in which the students live, it is possible to expect exploratory learning in accordance with the characteristics of the region, in collaboration and cooperation with local materials and human resources. In this way, looking at Japanese elementary education as a whole at a macro level, it will be important to continue the cycle of using STEAM education as a place and opportunity to practice the learning of each subject and realize the significance and usefulness of each subject through learning in STEAM education.

Conclusion: STEAM teaching strategies in elementary schools in Japan

In STEAM education, which aims at intercurricular collaboration or integration, attention tends to focus on perspectives such as how to collaborate and integrate. For example, Drake and Burns (2004) defined three levels of integration: multidisciplinary, which organizes standards of the disciplines around a theme; interdisciplinary, which organizes interdisciplinary skills and concepts embedded in disciplinary standards; and transdisciplinary, which organizes real-life contexts and student questions. However, as a prerequisite, it is also important to consider who performs the integration. Is it a pupil or a teacher? If pupils are integrated, it is preferable to integrate the knowledge and skills of each subject in secondary school, where the pupils' foundations for each subject have been established to a certain extent. However, if teacher integration is envisaged, an elementary school in which all subjects are taught by a teacher is preferable to a secondary school with a subject-teacher system. This may seem like a dilemma at first glance, but it means that STEAM teaching strategies can be clarified depending on who integrates them. Because this chapter focuses on STEAM teaching in elementary schools, "STEAM teaching strategies based on teacher integration" are discussed. The manner in which STEAM is perceived differs slightly depending on the position of the subject. It is extremely difficult for a person in a specialized subject to think from the perspective of another specialized subject; however, elementary school teachers, who teach all subjects, have the potential to overcome this barrier.

Acknowledgments

This work was supported by the Japan Society for the Promotion of Science KAKENHI Grant Numbers JP19K14344, JP20K20832, JP21H00919 (JP23K0744), JP22K03006.

Additional notes

This study has been significantly revised by adding new analyses based on the data presented at the international conferences of the European Science Education Research Association (ESERA 2023).

References

Akerson, V. L., Burgess, A., Gerber, A., Guo, M., Khan, T. A., & Newman, S. (2018). Disentangling the meaning of STEM: Implications for science and teacher education. *Journal of Science Teacher Education*, *29*(1), 1–8. https://doi.org/10.1080/1046560X.2018.1435063

Arima, A., Kobayashi, M., Hioki, M., Asai, M., Imamura, T., Umeda, T., Ouchi, T., Ohshika, K., Ohtaka, I., Ogawa, T., Ogawa, M., Kawakami, S., Kumano, Y., Kuroda, A., Kojima, T., Koshima, T., Goto, T., Goto, K., Kobayashi, M., ... Watanabe, J. (2020a). *Tanoshii Rika 3-nen (Grade 3 elementary science textbook – Fun science)*. Dainippon Tosho [in Japanese].

Arima, A., Kobayashi, M., Hioki, M., Asai, M., Imamura, T., Umeda, T., Ouchi, T., Ohshika, K., Ohtaka, I., Ogawa, T., Ogawa, M., Kawakami, S., Kumano, Y., Kuroda, A., Kojima, T., Koshima, T., Goto, T., Goto, K., Kobayashi, M., ... Watanabe, J. (2020b). *Tanoshii Rika 4-nen (Grade 4 elementary science textbook – Fun science)*. Dainippon Tosho [in Japanese].

Arima, A., Kobayashi, M., Hioki, M., Asai, M., Imamura, T., Umeda, T., Ouchi, T., Ohshika, K., Ohtaka, I., Ogawa, T., Ogawa, M., Kawakami, S., Kumano, Y., Kuroda, A., Kojima, T., Koshima, T., Goto, T., Goto, K., Kobayashi, M., ... Watanabe, J. (2020c). *Tanoshii Rika 5-nen (Grade 5 elementary science textbook – Fun science)*. Dainippon Tosho [in Japanese].

Arima, A., Kobayashi, M., Hioki, M., Asai, M., Imamura, T., Umeda, T., Ouchi, T., Ohshika, K., Ohtaka, I., Ogawa, T., Ogawa, M., Kawakami, S., Kumano, Y., Kuroda, A., Kojima, T., Koshima, T., Goto, T., Goto, K., Kobayashi, M., ... Watanabe, J. (2020d). *Tanoshii Rika 6-nen (Grade 6 elementary science textbook – Fun science)*. Dainippon Tosho [in Japanese].

Caiman, C., & Jakobson, B. (2022). Aesthetic experience and imagination in early elementary school science – A growth of 'Science–Art–Language–Game'. *International Journal of Science Education*, *44*(5), 833–853. https://doi.org/10.1080/09500693.2021.1976435

Drake, S. M., & Burns, R. C. (2004). *Meeting standards through integrated curriculum*. Association for Supervision and Curriculum Development (ASCD).

Fortus, D., Lin, J., Neumann, K., & Sadler, T. D. (2022). The role of affect in science literacy for all. *International Journal of Science Education*, *44*(4), 535–555. https://doi.org/10.1080/09500693.2022.2036384

Ishiura, S., Kamata, M., Ohsumi, Y., Kuryu, Y., Aoki, H., Akao, A., Akiyoshi, H., Abe, O., Arima, T., Anno, T., Ishikawa, S., Itou, A., Itonori, S., Irizuki, T., Uchiyama, H., Une, K., Etou, T., Ozaki, H., Osada, T., ... Watanabe, S. (2020a). *Wakuwaku Rika 3 (Grade 3 elementary science textbook – Exciting science)*. Keirinkan [in Japanese].

Ishiura, S., Kamata, M., Ohsumi, Y., Kuryu, Y., Aoki, H., Akao, A., Akiyoshi, H., Abe, O., Arima, T., Anno, T., Ishikawa, S., Itou, A., Itonori, S., Irizuki, T., Uchiyama, H., Une, K., Etou, T., Ozaki, H., Osada, T., ... Watanabe, S. (2020b). *Wakuwaku Rika 4 (Grade 4 elementary science textbook – Exciting science)*. Keirinkan [in Japanese].

Ishiura, S., Kamata, M., Ohsumi, Y., Kuryu, Y., Aoki, H., Akao, A., Akiyoshi, H., Abe, O., Arima, T., Anno, T., Ishikawa, S., Itou, A., Itonori, S., Irizuki, T., Uchiyama, H., Une, K., Etou, T., Ozaki, H., Osada, T., ... Watanabe, S. (2020c). *Wakuwaku Rika 5 (Grade 5 elementary science textbook – Exciting science)*. Keirinkan [in Japanese].

Ishiura, S., Kamata, M., Ohsumi, Y., Kuryu, Y., Aoki, H., Akao, A., Akiyoshi, H., Abe, O., Arima, T., Anno, T., Ishikawa, S., Itou, A., Itonori, S., Irizuki, T., Uchiyama, H., Une, K., Etou, T., Ozaki, H., Osada, T., ... Watanabe, S. (2020d). *Wakuwaku Rika 6 (Grade 6 elementary science textbook – Exciting science)*. Keirinkan [in Japanese].

Isozaki, T., & Nozoe, S. (2017). Takuetsusei no kagaku kyoiku wo itoshita karikyuramu no kousei genri jyosetsu (A proposal on the organizing principles of science curricula aiming for "Science for Excellence"). *Journal of Science Education in Japan*, *41*(4), 388–397 [in Japanese].

Jolly, A. (2017). *STEM by design: Strategies and activities for grades 4–8*. Routledge.

Ministry of Education, Culture, Sports, Science and Technology (MEXT). (2021). *Promotion of cross-curricular learning such as STEAM education* [in Japanese]. https://www.mext.go.jp/content/20210716-mxt_kyoiku01-000016739_1.pdf

Mouri, M., Oshima, M., Amamiya, T., Andou, H., Iokawa, Y., Ishikawa, Y., Ishida, Y., Isono, I., Iwamoto, T., Ueki, T., Ohiyama, T., Ohki, S., Okada, T., Ogura, Y., Ono, N., Kasahara, M., Kataoka, S., Kato, T., Kato, N., ... Watabe, T. (2020a). *Atarashii Rika 3-nen (Grade 3 elementary science textbook – New science)*. Tokyo Shoseki [in Japanese].

Mouri, M., Oshima, M., Amamiya, T., Andou, H., Iokawa, Y., Ishikawa, Y., Ishida, Y., Isono, I., Iwamoto, T., Ueki, T., Ohiyama, T., Ohki, S., Okada, T., Ogura, Y., Ono, N., Kasahara, M., Kataoka, S., Kato, T., Kato, N., ... Watabe, T. (2020b). *Atarashii Rika 4-nen (Grade 4 elementary science textbook – New science)*. Tokyo Shoseki [in Japanese].

Mouri, M., Oshima, M., Amamiya, T., Andou, H., Iokawa, Y., Ishikawa, Y., Ishida, Y., Isono, I., Iwamoto, T., Ueki, T., Ohiyama, T., Ohki, S., Okada, T., Ogura, Y., Ono, N., Kasahara, M., Kataoka, S., Kato, T., Kato, N., ... Watabe, T. (2020c). *Atarashii Rika 5-nen (Grade 5 elementary science textbook – New science)*. Tokyo Shoseki [in Japanese].

Mouri, M., Oshima, M., Amamiya, T., Andou, H., Iokawa, Y., Ishikawa, Y., Ishida, Y., Isono, I., Iwamoto, T., Ueki, T., Ohiyama, T., Ohki, S., Okada, T., Ogura, Y., Ono, N., Kasahara, M., Kataoka, S., Kato, T., Kato, N., ... Watabe, T. (2020d). *Atarashii Rika 6-nen (Grade 6 elementary science textbook – New science)*. Tokyo Shoseki [in Japanese].

Nesmith, S. M., & Cooper, S. (2020). Elementary STEM learning. In C. C. Johnson, M. J. Mohr-Schroeder, T. J. Moore, & L. D. English (Eds.), *Handbook of research on STEM education* (pp. 101–114). Routledge.

Nozoe, S. (2015). Katsuyogata rikajyugyo no kyouzai kenkyu no shiten (Perspectives on researching teaching materials for application-oriented science lessons). *Elementary School Science Education*, *48*(9), 7–10 [in Japanese].

Nozoe, S. (2019). Korekara no jidai ni taioushita atarashii rika no mondaikaiketsu – Wagakuni no dento to kokusaiteki na kagakukyoiku no cyouryu wo humaete (New science problem-solving in Japan for the future: Based on our traditions and international trends in science education). *Science Education Monthly*, *68*(6), 44–46 [in Japanese].

Shimoda, K., Morimoto, S., Akiyama, Y., Azuma, T., Isozaki, T., Inagaki, S., Imaizumi, T., Urano, K., Ohnishi, H., Odagiri, M., Kai, H., Kamata, M., Kiryu, T., Kubota, Y., Kurihara, J., Kobayashi, H., Sato, K., Sato, T., Sasaki, A., ... Watanabe, M. (2020a). *Minna to Manabu Syougakko Rika 3-nen (Grade 3 elementary science textbook – Science with everyone)*. Gakkotosho [in Japanese].

Shimoda, K., Morimoto, S., Akiyama, Y., Azuma, T., Isozaki, T., Inagaki, S., Imaizumi, T., Urano, K., Ohnishi, H., Odagiri, M., Kai, H., Kamata, M., Kiryu, T., Kubota, Y., Kurihara, J., Kobayashi, H., Sato, K., Sato, T., Sasaki, A., ... Watanabe, M. (2020b). *Minna to Manabu Syougakko Rika 4-nen (Grade 4 elementary science textbook – Science with everyone)*. Gakkotosho [in Japanese].

Shimoda, K., Morimoto, S., Akiyama, Y., Azuma, T., Isozaki, T., Inagaki, S., Imaizumi, T., Urano, K., Ohnishi, H., Odagiri, M., Kai, H., Kamata, M., Kiryu, T., Kubota, Y., Kurihara, J., Kobayashi, H., Sato, K., Sato, T., Sasaki, A., ... Watanabe, M. (2020c). *Minna to Manabu Syougakko Rika 5-nen (Grade 5 elementary science textbook – Science with everyone)*. Gakkotosho [in Japanese].

Shimoda, K., Morimoto, S., Akiyama, Y., Azuma, T., Isozaki, T., Inagaki, S., Imaizumi, T., Urano, K., Ohnishi, H., Odagiri, M., Kai, H., Kamata, M., Kiryu, T., Kubota, Y., Kurihara, J., Kobayashi, H., Sato, K., Sato, T., Sasaki, A., ... Watanabe, M. (2020d). *Minna to Manabu Syougakko Rika 6-nen (Grade 6 elementary science textbook – Science with everyone)*. Gakkotosho [in Japanese].

Takeuchi, M. A., Sengupta, P., Shanahan, M.-C., Adams, J. D., & Hachem, M. (2020). Transdisciplinarity in STEM education: A critical review. *Studies in Science Education*, *56*(2), 213–253.

Tyrrell, N. (2020a). *Foxton primary science: All about plants (lower KS2 science)*. Foxton Books.

Tyrrell, N. (2020b). *Foxton primary science: Electricity (lower KS2 science)*. Foxton Books.

Tyrrell, N. (2020c). *Foxton primary science: Light (lower KS2 science)*. Foxton Books.

Tyrrell, N. (2020d). *Foxton primary science: Living things and their changing habitats (lower KS2 science)*. Foxton Books.

Tyrrell, N. (2020e). *Foxton primary science: Magnets and friction (lower KS2 science)*. Foxton Books.

Tyrrell, N. (2020f). *Foxton primary science: Predators and prey (lower KS2 science)*. Foxton Books.

Tyrrell, N. (2020g). *Foxton primary science: Rocks (lower KS2 science)*. Foxton Books.

Tyrrell, N. (2020h). *Foxton primary science: Sound (lower KS2 science)*. Foxton Books.

Tyrrell, N. (2020i). *Foxton primary science: States of matter solids, liquids and gases (lower KS2 science)*. Foxton Books.

Tyrrell, N. (2020j). *Foxton primary science: The human body (lower KS2 science)*. Foxton Books.

Tyrrell, N. (2020k). *Foxton primary science: Classification (upper KS2 science)*. Foxton Books.

Tyrrell, N. (2020l). *Foxton primary science: Earth and space (upper KS2 science)*. Foxton Books.

Tyrrell, N. (2020m). *Foxton primary science: Electricity (upper KS2 science)*. Foxton Books.

Tyrrell, N. (2020n). *Foxton primary science: Evolution and inheritance (upper KS2 science)*. Foxton Books.

Tyrrell, N. (2020o). *Foxton primary science: Forces (upper KS2 science)*. Foxton Books.

Tyrrell, N. (2020p). *Foxton primary science: Life cycles and reproduction (upper KS2 science)*. Foxton Books.

Tyrrell, N. (2020q). *Foxton primary science: Light (upper KS2 science)*. Foxton Books.

Tyrrell, N. (2020r). *Foxton primary science: Prehistoric life (upper KS2 science)*. Foxton Books.

Tyrrell, N. (2020s). *Foxton primary science: Properties and changes of materials (upper KS2 science)*. Foxton Books.

Tyrrell, N. (2020t). *Foxton primary science: The human body (upper KS2 science)*. Foxton Books.

Wong, V., Dillon, J., & King, H. (2016). STEM in England: Meanings and motivations in the policy arena. *International Journal of Science Education, 38*(15), 2346–2366. https://doi.org/10.1080/09500693.2016.1242818

Yoro, T., Kadoya, S., Maruyama, S., Aiba, H., Ishii, M., Itaki, T., Itaba, O., Itou, T., Inada, Y., Iwasaki, Y., Oikawa, S., Ohnuki, A., Ogawa, H., Katahira, K., Kawasaki, K., Kitamura, K., Kinoshita, H., Sakai, T., Sakamoto H., ... Watanabe, H. (2020a). *Mirai wo Hiraku Syougakko Rika 3 (Grade 3 elementary science textbook – Science for the future)*. Kyoiku Shuppan [in Japanese].

Yoro, T., Kadoya, S., Maruyama, S., Aiba, H., Ishii, M., Itaki, T., Itaba, O., Itou, T., Inada, Y., Iwasaki, Y., Oikawa, S., Ohnuki, A., Ogawa, H., Katahira, K., Kawasaki, K., Kitamura, K., Kinoshita, H., Sakai, T., Sakamoto H., ... Watanabe, H. (2020b). *Mirai wo Hiraku Syougakko Rika 4 (Grade 4 elementary science textbook – Science for the future)*. Kyoiku Shuppan [in Japanese].

Yoro, T., Kadoya, S., Maruyama, S., Aiba, H., Ishii, M., Itaki, T., Itaba, O., Itou, T., Inada, Y., Iwasaki, Y., Oikawa, S., Ohnuki, A., Ogawa, H., Katahira, K., Kawasaki, K., Kitamura, K., Kinoshita, H., Sakai, T., Sakamoto H., ... Watanabe, H. (2020c). *Mirai wo Hiraku Syougakko Rika 5 (Grade 5 elementary science textbook – Science for the future)*. Kyoiku Shuppan [in Japanese].

Yoro, T., Kadoya, S., Maruyama, S., Aiba, H., Ishii, M., Itaki, T., Itaba, O., Itou, T., Inada, Y., Iwasaki, Y., Oikawa, S., Ohnuki, A., Ogawa, H., Katahira, K., Kawasaki, K., Kitamura, K., Kinoshita, H., Sakai, T., Sakamoto H., ... Watanabe, H. (2020d). *Mirai wo Hiraku Syougakko Rika 6 (Grade 6 elementary science textbook – Science for the future)*. Kyoiku Shuppan [in Japanese].

Zeidler, D. L. (2016). STEM education: A deficit framework for the twenty first century? A sociocultural socioscientific response. *Cultural Studies of Science Education, 11*(1), 11–26. https://doi.org/10.1007/s11422-014-9578-z

5 Assessment for fostering Japanese STEM literacy

From a perspective of performance assessment theory

Terumasa Ishii

Introduction

This chapter examines the Japanese approach to STEM literacy assessment, based on the accumulation of performance assessment theories in Japan. STEM/STEAM education has been pursued in Japan in science, mathematics, technology, and other subject areas, as well as in "period for integrated studies." In particular, as competency-based curriculum reforms have developed, STEM/STEAM education has been implemented as cross-curricular learning and inquiry-based learning, in which knowledge and skills are used in a multidisciplinary manner in authentic real-world contexts. As a method for assessing such authentic and meaningful learning, the findings of the performance assessment (PA) theory have been referred to. In this chapter, I summarize the development of the theory and practice utilizing PA in science, mathematics, technology, and other subjects, as well as in period for integrated studies, and clarify the ideal form of assessment for fostering STEM literacy.

When considering assessment, it is necessary to clarify the goals of what we want to foster in STEM/STEAM education, which corresponds to the concept of "STEM literacy" in this book, however, to begin with, STEM/STEAM education is not a curriculum area with clear content and structure. STEM/STEAM education is not a curriculum area but rather a slogan, and the meaning of the term depends on the context of educational reforms in each country. In this chapter, we first clarify the characteristics of STEM/STEAM education in Japan in terms of how it is perceived, developed, and practiced, and what educational principles and values are embedded in the term. In particular, STEM/STEAM education in Japan is developing within the magnetic field of the Japanese version of competency-based educational reform, a reform based on "quality and ability," as described below.

This chapter therefore examines the characteristics of competency-based reform in Japan and discusses the features of STEM/STEAM education in Japan and what is being aimed for there. The chapter then outlines the basic concepts and methods of performance assessment, and proposes a comprehensive framework for capturing the quality and ability (learning outcomes) that STEM/STEAM education in Japan fosters, considering that it is practiced in

DOI: 10.4324/9781003392545-6

a holistic educational manner, spanning both academic subjects and integrated learning, and including the cultivation of generic skills. Finally, I propose a comprehensive framework that captures the quality and ability (learning outcomes) to be fostered and discuss how STEM/STEAM education should be assessed by presenting examples of subject-based, cross-curricular, period for integrated studies, and project study.

Position of STEM/STEAM education in Japan—a magnetic field of quality—and ability-based education

Development of competency-based reforms

In response to the demands of modern society for human resource development, especially from industry, which has been described as late modernity, post-modernity, and high modernity, a trend (competency-based educational reform) to emphasize the development of competencies (the ability to predict professional competence and success in life) has developed worldwide (Matsushita, 2010; Rychen & Salganik, 2003).

In the 2000s, inspired by the OECD's PISA and key competencies, Japan, spanning primary, secondary, and higher education, emphasized quality and ability, generic skills, as well as learner-centered, cross-disciplinary, and active learning, and moved from content-based to competency-based (quality and ability) curriculum reform. In Japan, this has developed as a reform based on child-centeredness and progressivism (Ishii, 2022). The progressivist narrative emphasizes learner-centeredness and the ability to learn and proactive attitudes (process), rather than focusing on teacher-leadness and the content of learning (outcome), and was reinforced by the development of constructivist learning psychology, which views knowledge as constructed by learners and emphasizes the active nature of learners, providing scientific evidence. In addition, in educational discourses, the language of psychology and cognitive and learning science, such as collaborative learning, knowledge construction, conceptual knowledge, thinking skills, metacognition, self-regulation, and socioemotional skills, was expanded. In contrast, with the use of the neutral term "quality and ability," the orientation toward reconstructing subject content from the perspective of meaning and relevance to society and individuals, which was inherent in the concept of competence, receded to the background.

In addition, with the nationwide installation of one terminal per student triggered by the COVID-19 pandemic, the importance of individualized and personalized learning has been raised, and reforms based on quality and ability have been relegated to a discourse emphasizing proactive attitudes, and there is a strong tendency toward abstract skills and attitudes (inner mind) that lack objectivity rather than toward concrete society and the world (outer world). Thus, the explicit goal of fostering talented human resources in the economic world has been avoided. However, the human image is abstracted through psychological and neutral discourses, and differences in perspective and value conflicts about what kind of society to aim for are not revealed. This results in a convergence of human development that is adaptive to society.

Various layers of STEM/STEAM education in Japan

The reform of the Japanese education system based on quality and ability described above steers toward a progressive educational orientation and psychologization of education that aims for learner-centered, cross-disciplinary, and borderless learning beyond the framework of subjects and school settings created by adult society. This magnetic field also characterizes STEM/STEAM education in Japan.

The context in which STEM education has emerged in the United States is the changing industrial structure of the 1990s and the strengthening of international competitiveness, which is commonly associated with competency-based reforms. In response to the specific social changes of the late modern era, as conceptualized in terms such as the "knowledge economy" and the "fourth industrial revolution," the challenge became the development of scientific and technological human resources and the cultivation of scientifically literate citizens. While it was stated that the subjects should be integrated rather than separate and that they should lead to the development of 21st-century skills, what was directly aimed at was the interconnectedness of science, technology, engineering, and mathematics, and the social relevance of each subject and discipline. Historically, it can be viewed as a movement for the renewal of science education curriculum in a broad sense, following the controversy over science and culture in the late 19th century, the movement to remodel mathematics education in the early 20th century, and the movement to modernize subject content in the 1960s (Isozaki & Isozaki, 2021).

In Japan, STEM education was initially introduced with an orientation toward significant learning activities that integrate the use of knowledge and skills across subjects, with a focus on science education fields and incorporating practices that relate to society and the world, as well as an orientation to question the nature of science education in relation to engineering and other fields (Isobe & Yamazaki, 2015; Kumano, 2014). However, in Japan, the reforms based on quality and ability tend to be formally regarded as an emphasis on generic skills, and the period for integrated studies, an area for developing cross-cutting and integrated learning, is explicitly and systematically positioned in the curriculum. The emphasis tended to be more on the cross-curricular nature of STEM education, rather than on the content renewal aspect of science education.

For example, Matsubara and Kosaka (2017) summarized research trends in STEM education in Japan and stated that there was a growing trend to view STEM in an integrated manner. They then referred to Vasquez et al. (2013) to classify STEM education by the degree of curriculum integration. The characteristics of quality and ability to be fostered in STEM education are summarized in Table 5.1.

Table 5.1 Degree of integration and the qualities and abilities to be developed in particular (Matsubara & Kosaka, 2017, p. 157).

Degree of integration	approach	Qualities and abilities to be developed in particular
(Differentiation)		
Low	Thematic or Multidisciplinary	Subject-specific concepts and individual skills
	Interdisciplinary	Cross-curricular concepts and generic skills
High	Transdisciplinary	Ability to solve real-world problems
(Integration)		

Matsubara (2020) stated, "STEM/STEAM education generally emphasizes problem-solving and project-based learning due to the expectation of fostering innovation and creativity" (p. 10), and that the difference from other project-based learning is a strong awareness of the specificity of each subject and domain. He also sees the learning process as one that "tackles authentic issues using ideas specific to each subject and domain" (pp. 10–11).

Furthermore, the "Future Classrooms" initiative by the Ministry of Economy, Trade, and Industry (Ministry of Economy, Trade and Industry [METI], 2019) was the catalyst for the increased attention to the keyword "STEAM" in Japan. The "Future Classrooms" concept set forth the "STEAM-ization of learning" as a pillar for fostering changemakers who will create the future. It is expected that students will acquire subject knowledge and expertise (i.e., "know") regardless of their background in the humanities and sciences, will think creatively and logically by connecting knowledge horizontally through inquiry- and project-based learning (PBL), and will be able to solve problems. To promote these efforts, the use of EdTech, developed by private companies and others, is recommended. Although elements of liberal arts (arts) were added to "STEAM-ization of learning," it is in effect reinterpreted as design thinking rather than a liberal arts or humanities orientation, and is intended to support industrial competitiveness, such as manufacturing using robotics and programming, and deepening understanding of science and technology through experimentation and experience. The STEAM library was also established to promote the development of STEAM learning programs and their digital content through industry–academia–government collaboration.

In the "Future Classrooms," STEAM education is more connected to inquiry-based learning, and the emphasis is placed on the thinking and inquiry process itself, which cycles between "knowing" and "creating," rather than on the specifics of each subject or area. The themes handled are oriented to emphasize children's excitement and fun, including students-led school rule-making. The themes are also broadly defined (Asano, 2022). When the term "STEM" education was used, it mainly included aspects such as an interdisciplinary and cross-disciplinary learning based on confirmation of the expertise and essentiality of each subject and field, and as a movement for renewal of educational content in general education (updating social and world awareness, which is the basis of the "what to know" debate). However, when STEAM education is used, it is strongly oriented toward human resource development and tends to be psychologically abstracted as content-free, inquiry-mode learning (cultivating internal attitudes).

Focus on performance assessment in Japan's educational reform

Basic concept and methods of performance assessment

As described above, STEM/STEAM education in Japan is being developed in a wide range of fields and learning settings, including some subjects, period for integrated studies, and *tokkatsu* (student-led activities). The forms of STEM/STEAM education vary from subject-specific to integrated, however, they all share a common orientation toward fostering skills and attitudes that go beyond subject content mastery in authentic contexts. In addition, in the assessment of such quality and ability, PA has been the focus (Nishioka, 2016). PA is generally a qualitative method of assessing learners' ability to understand the meaning of concepts and to use knowledge and skills comprehensively, based on their behavior and works (performance) produced in the authentic and real contexts where thinking is necessary (Wiggins & McTighe, 2005). In a narrow sense, it means "performance task-based assessment" in which performance tasks (PTs) are designed to assess learners' abilities, such as setting realistic and meaningful scenarios, and the process and products of activities in response to these scenarios are assessed. Examples of PT include a music class project to add background music to a school introduction video, a science class project to design and propose a blueprint (electric circuit) for an electric car, and a social studies class project to devise an action plan to reduce litter on the beach and propose it to the local community. In a broader sense, PA also refers to "performance-based assessment," such as informal and formative assessment of the process of children's daily learning activities based on what they say, do, and write in their notebooks during class. Portfolio assessment methods are another type of PA.

Design of assessment methods according to the qualitative level of learning outcomes

To design assessments that live up to instructional improvement, it is necessary to clarify the target learning outcomes and design suitable assessment methods. For example, the cognitive domain of Bloom's Taxonomy comprises six major categories: (1) knowledge; (2) comprehension; (3) application;

(4) analysis; (5) synthesis; and (6) evaluation; whose hierarchical structure indicates the qualitative level of quality, ability, and learning activities. The hierarchy of the taxonomy of cognitive domains is then largely captured in three layers: (1) knowledge (memorizing of factual knowledge); (2) comprehension (of conceptual knowledge); and application (of procedural knowledge) (3) "analysis," "synthesis," and "evaluation" (higher-order problem solving that combines various types of knowledge according to the situation).

Guidelines for the design of assessment methods according to the qualitative level of target learning outcomes can be summarized as follows. The "reproducing" level that questions the acquisition of discrete knowledge and skills (e.g., answering terms such as "population" and "sample mean") can be assessed by objective tests such as fill-in-the-blank questions and choice-type questions. However, at the "understanding" level, which questions understanding of a concept (e.g., "the quality of sweets manufactured by a certain food company" is presented as a research situation, and the student must decide whether a total survey or a sample survey is more appropriate and answer why), connections and images among knowledge are important. It is not possible to make a judgment without opportunities such as asking students to explain a certain concept with examples, having them express the structure and images in their minds on a picture or mind map, or having them solve application problems. Furthermore, the "practicing" level, which questions the ability to comprehensively apply knowledge and skills in the context of real life and society (e.g., creating a research plan to estimate the number of light vehicles in Hiroshima City), cannot be assessed without actually having the students express, create, or try. In this way, PT is a task to assess the ability (competence) to do so by having students try out practices that involve thinking.

In recent years, in the global development of competency-based curriculum reform, generic skills, etc., have also been conceptualized. However, research on taxonomies of educational objectives in Western education, which is characterized as an intellectual school, has been conducted primarily with subject learning in mind. In contrast, an extension of the taxonomy is necessary to capture the whole picture of quality and ability (learning outcomes) that can be nurtured by the curriculum in Japanese schools, where not only academic subjects but also integrated learning, student-led activities, etc. are systematically positioned as part of the official curriculum (Ishii, 2017).

As shown in Figure 5.1, in subject learning that are organized in the form of content units and in which the cognitive system is the main target of instruction, the teacher sets the general framework of the content and learning tasks, but in period for integrated studies and the like, which are organized in the form of thematic units. The task setting is often left to the children (metacognitive system). Furthermore, in extra-curricular activities such as *tokkatsu* (student-led activities), the relationships and rules (context) of the learning community itself are jointly reconstructed or newly constructed by the children (action system). In period for integrated studies, which is developed based on PBL, children discover issues on their own and continue to learn in a sustained and

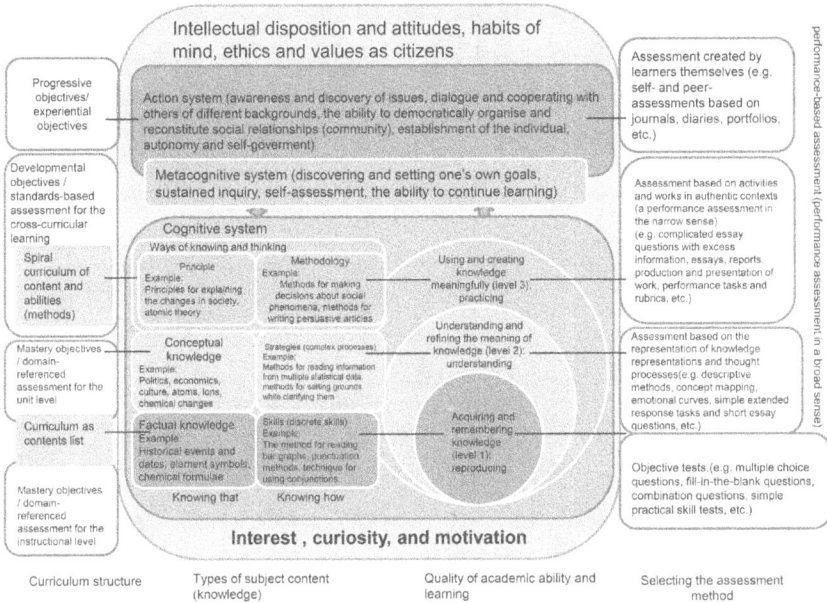

Intellectual disposition and attitudes, habits of mind, ethics and values as citizens

Progressive objectives/ experiential objectives

Action system (awareness and discovery of issues, dialogue and cooperating with others of different backgrounds, the ability to democratically organise and reconstitute social relationships (community), establishment of the individual, autonomy and self-goverment)

Assessment created by learners themselves (e.g. self- and peer-assessments based on journals, diaries, portfolios, etc.)

Developmental objectives / standards-based assessment for the cross-curricular learning

Metacognitive system (discovering and setting one's own goals, sustained inquiry, self-assessment, the ability to continue learning)

Spiral curriculum of content and abilities (methods)

Cognitive system
Ways of knowing and thinking

Principle
Example:
Principles for explaining the changes in society, atomic theory

Methodology
Example:
Methods for making decisions about social phenomena, methods for writing persuasive articles

Using and creating knowledge meaningfully (level 3): practicing

Assessment based on activities and works in authentic contexts (a performance assessment in the narrow sense) (e.g. complicated essay questions, essays, reports, production and presentation of work, performance tasks and rubrics, etc.)

Mastery objectives / domain-referenced assessment for the unit level

Conceptual knowledge
Example:
Politics, economics, culture, atoms, ions, chemical changes

Strategies (complex processes)
Example:
Methods for reading information from multiple statistical data, methods for setting grounds while clarifying them

Understanding and refining the meaning of knowledge (level 2): understanding

Assessment based on the representation of knowledge representations and thought processes(e.g. descriptive methods, concept mapping, emotional curves, simple extended response tasks and short essay questions, etc.)

Curriculum as contents list

Factual knowledge
Example:
Historical events and dates, element symbols, chemical formulae

Skills (discrete skills)
Example:
The method for reading bar graphs, punctuation methods, technique for using conjunctions

Acquiring and remembering knowledge (level 1): reproducing

Mastery objectives / domain-referenced assessment for the instructional level

Knowing that   Knowing how

Objective tests (e.g. multiple choice questions, fill-in-the-blank questions, combination questions, simple practical skill tests, etc.)

Interest , curiosity, and motivation

Curriculum structure   Types of subject content (knowledge)   Quality of academic ability and learning   Selecting the assessment method

performance-based assessment (performance assessment in a broad sense)

Figure 5.1 Framework for grasping the hierarchical nature (qualitative level) of capabilities nurtured in schools.

Source: Ishii (2017, p. 47). The framework for articulating the quality of academic ability and learning is a slightly modified version of R.J. Marzano's Dimensions of Learning framework (Marzano, 1992), while the typology of subject content is a reworking of G. Wiggins's Structure of Knowledge framework (Wiggins & McTighe, 2005).

collaborative manner. To assess long-lasting efforts, reports and works produced through PBL, the process of thinking and communication that produce the reports and works, and the reflections on the learning process (experienced curriculum) are assessed. In period for integrated studies, not only the works (reports, papers, proposals, social actions, speeches, physical expressions, plays, aesthetic works, etc.) produced through activities that deepen inquiry into particulars, research questions, social issues, etc. but also the individualized inquiry process of each child is assessed based on their trial and error. It is important to discover and support the value of learning in the process of trial-and-error, so the use of the portfolio assessment method is also effective.

Quality and ability to be assessed in period for integrated studies

Table 5.2 shows examples of the primary knowledge, skills, and dispositions (elements of quality and ability) required for each hierarchical level of learning outcomes. As in the aforementioned "Future Classrooms" concept, there is often a dualistic theory of stages or division of responsibilities, such as acquiring knowledge in academic subjects and developing the

Table 5.2 A framework for capturing the overall picture of the elements of qualities and abilities to be nurtured in schools (Ishii, 2017, p. 48).

Hierarchical level of abilities and learning activities (structure of the curriculum)		Elements of qualities and abilities (pillars of objectives)			
		Knowledge	Skills		Affective traits (interest, motivation, attitude, personality characteristics)
			Cognitive skills	Social skills	
Subject instruction	Learning within the contents, tasks, and context given by teachers				
	Acquiring and remembering knowledge (reproducing)	Factual knowledge, skills (discrete skills)	Memorizing and reproduction, mechanical execution and automation	Learning from each other, constructing knowledge jointly	Feeling a sense of self-efficacy due to achieving one's goals
	Understanding and refining the meaning of knowledge (understanding)	Conceptual knowledge, strategies (complex processes)	Interpreting, relating, structuring, comparing and classifying, deductive and inductive reasoning, generalizing and specifying		Intrinsic motivation in accordance with the value of the content, interest, and desire for the content
	Using and creating knowledge meaningfully (practicing)	A complex of domain-specific knowledge, with a focus on ways of knowing and thinking (principles and methodology)	Intelligent problem solving; decision making; proving, experimenting, and investigating including abductive inference; discovery and invention of knowledge and objects; aesthetic expression (critical thinking and creative thinking are involved)	Dialogue and communication to conduct projects and working cooperatively	Intrinsic motivation in accordance with the social relevance of activities, belief about subject matter and learning (intellectual disposition and attitudes, and habits of mind)

(Continued)

Table 5.2 (Continued)

Hierarchical level of abilities and learning activities (structure of the curriculum)	Elements of qualities and abilities (pillars of objectives)			
	Knowledge	Skills		Affective traits (interest, motivation, attitude, personality characteristics)
		Cognitive skills	Social skills	
Integrated learning — Students themselves are able to decide and re-structure the learning context and conditions — Autonomous sustained inquiry (metacognitive system)	Thoughts and views, world view and self-image	Discovering and setting one's own goals, sustained inquiry, information collection and processing, self-assessment		Intrinsic motivation rooted in one's own thoughts and motivation for life (earnestness), and the formation of aspirations and career awareness
Student-led activities — Autonomous organizing and reconstituting social relationships (action system)	Awareness of relationships between people and the community and culture to which one belongs, methodology related to running the community and maintaining its autonomy	Solving life problems, formulating events and plans, involvement and participation in planning how to resolve social problems	Human relationships and exchanges (teamwork), rules and division of labor, leadership and management, handling of disputes and reaching agreement, independent organizing and reconstituting of the sites and community of learning	Social motivation rooted in social responsibility and moral consciousness, establishment of an ethical life philosophy and positionality

Notes:
* The classification of the level contains a dotted line in the sections for social skills and affective traits; this indicates that the correspondence relationship for each level is loose, compared to knowledge and cognitive skills.
* The highlighted portions are the elements of the objectives that the curriculum clearly indicates. One should mainly be conscious of these aspects in regards to the respective levels of abilities and learning activities.
* The content of cognitive and social skills should be tailored to each school. The cases indicated in the National Courses of Study should be used as reference materials. In the field of affective traits, the goal should be to build a formative or curriculum assessment, rather than establish a rating.

ability to think and the attitudes through inquiry in period for integrated studies. However, as shown in Table 5.2, any learning activity involves the development of some knowledge, skills, and attitudes. Therefore, it is necessary to clarify which quality of thinking is aimed for in each of the subjects and integrated learning. If we consider the unique meaning of period for integrated studies in the curriculum, students need to experience the process of setting and reconstructing their own issues and questions that are not resolved in thinking that is in line with the content and methodology of the subjects and that can be used to master them. It should also focus on the experience of collecting and analyzing necessary information on one's own.

Thus, while being aware of the qualitative level of learning outcomes and keeping in mind the cycle of inquiry-based learning activities, we can clarify the main elements of quality and ability to be nurtured, such as problem setting, information gathering and analysis, the ability to accomplish tasks, collaboration, self-assessment, and expression, and from these perspectives, we can describe and assess the qualitative changes and growth in children's performance. The experienced children's learning history accumulated in portfolios, etc., can be judged in terms of specific goals, perspectives, and assessment criteria, and rubrics can be used as a guide for teachers' assessment and children's self-assessment, while the level of ability, etc., can also be judged per the goals. In contrast, it is important not only to develop children's abilities to a certain level, but also to describe and interpret the unique value of each child's experience and narrative in an intra-personal, evaluative manner, recognizing the child's individuality, strengths, and progress. While adhering to the goals, the teacher must be sensitive and discerning enough to recognize the value of the child's learning and experiences that are not limited to those goals.

STEM/STEAM education and assessment in subject learning and integrated learning

STEAM-like performance assessment in science, mathematics, and technology

Now, we will present some issues and points related to the nature of assessment while introducing examples of assessment practices related to STEM/STEAM education. STEM/STEAM assessments in other countries are generally conducted as assessments within each subject area (Suzuki, 2022). Oshima (2022), as a suggestion for Japan based on the assessment trend of STEM/STEAM education in Singapore, stated:

> When introducing STEM education, it is not necessary to establish a detailed STEM-specific assessment system … It is possible to view the purpose of introducing STEM as being for authentic learning or assessment of each subject and to use the assessment framework in existing subjects such as science to assess only those areas where possible.
>
> (p. 100)

Thus, STEM/STEAM education that values the specificity of each subject is considered an effort to question the nature of learning in existing subjects in the direction of emphasizing social relevance and authentic learning. This is where the use of performance tasks and rubrics that foster practicing-level academic abilities can be effective.

In considering the scene of authentic issues and identifying the essential questions and processes of the subject matter, it is important to consider both the social and practical context of citizens, workers, and consumers and the academic and cultural context as professional inquiry of researchers or as hobby and culture of the general public (Ishii, 2020). The social and practical context r is to reconsider the significance of learning a subject from the perspective of learning the subject as "glasses" to read the real world. For example, a function, from its origin, is a tool for predicting the future, and if one can master probability, it can help predict and judge risks by quantitatively grasping trends. In contrast, in terms of the academic and cultural context, the fun of academic study, it is important to ask, "Why is it right? Is it true in every situation?" It is important to ask whether students can experience the authentic process of "knowledge creation" in each subject and field of study, that is, whether they can delve into premises, logically reason based on what has already been clarified, and develop new conclusions and questions.

The pursuit of authentic learning in a subject is to be understood as the creation of "do a subject" classes (classes in which students deepen the essence of the subject together by experiencing situations in which knowledge and skills are used in real life or the process of knowledge exploration by experts in the field), as opposed to "learn a subject" classes. STEM/STEAM education, while fundamentally focusing on social and practical contexts, will ensure its quality and dynamic nature by interacting with academic and cultural contexts.

Figure 5.2 is a science performance assignment. Although it is a simulation of a real-world context, the design of the assignment considers essential questions and enduring understandings, and the content outline is relatively clear, such as the identification of substances using the nature of polymeric compounds, which are being tested. However, it is not a simple thought process of simply applying specific content to solve a problem, but a long and complex "practicing" level of thinking between question and answer to formulate the problem situation and synthesize existing knowledge and skills. The rubric assesses the process of scientific thinking, such as the planning of the experimental design and the validity of the decisions based on the results, rather than conceptual understanding of polymeric compounds.

In addition, Hisashi Miura (former teacher at Junior High School Attached to Faculty of Education, Kumamoto University) presented a unit on solving problems by programming about interactive content using networks in the second-year junior high school technology and home economics class, in which students were asked to evaluate new technologies from multiple perspectives and think about how to improve their lives now or in the future. The students work on the task "Let's propose contents that can improve energy

PT:

You are a researcher for a pharmaceutical company. You are to investigate the active ingredients of an antipyretic drug from another company and the additives contained in the drug, and report your findings to your supervisor. Each of you (in your group) is to plan and conduct the experiment and prepare a report.

Essential Question:

How can organic and (natural) macromolecular compounds in everyday substances be identified?

Enduring understanding:

To identify organic and (natural) macromolecular compounds around us, we can use coloring reactions and the like.

stage	Experimental planning, experimental, and observational skills	Thinking, Judgment, and Expression
Sperior	The student is able to plan experiments appropriately and conduct experiments safely and accurately. The student is able to accurately observe and record the results of experiments.	The student understands the meaning of the experiment and correctly identify antipyretic agents and additives from the results.
A	The student is able to plan experiments appropriately and conduct experiments safely. The student is able to observe and record the results of experiments.	The student understands the meaning of the experiment and correctly identify additives from the results.
B	The student is able to plan experiments appropriately and conduct experiments safely. The student is able to observe and record the results of experiments but is somewhat inadequate in observing and recording the results.	The student understands the meaning of the experiment and correctly identify the antipyretic from the results.
C	Lack of planning in experiments. The student is able to conduct experiments, but poor observation and recording.	The student does not understand the meaning of the experiment.

Figure 5.2 PT of "Identification of antipyretic agents and additives in medicines" in High School Chemistry (Unit "Properties and Utilization of Polymer Compounds") (Fukumoto, 2020).

conservation and infection control problems at home." Students reviewed their surroundings, noticed problems, and envisioned solutions using technology. They developed content individually and in groups, envisioning realistic practical situations, such as a program that automatically opens and closes doors based on the number of people in the room or body temperature, to avoid airtight spaces. They then created a YouTube video to introduce the content. Thus, PBL is organized with a great deal of room for student self-determination, including assignments, etc., and the process of setting and solving problems is assessed using a rubric. Next, let's look at an example of a cross-sectional, integrated STEM/STEAM education performance assessment.

Cross-sectional and integrated performance assessment during period for integrated studies, project study, etc.

As an example of cross-curricular learning, I introduce a case study of Fukuoka Primary School attached to University of Teacher Education Fukuoka (2023), which implements a cross-curricular, interdisciplinary, and interrelated curriculum that reexamines the conventional framework of subjects and values the context of children's learning. For example, Daisuke Nishijima's 6th grade mathematics unit, "Let's Explore the Secrets of Beauty of Family Crests," was related to art study and approached the question, "What is the secret of beautiful shapes?," which was generated from the cross-curricular theme of "What we want to leave behind" (exploration of tradition and culture). Specifically, the mathematical question "What kind of shape is characteristic of a beautiful pattern?" was the axis of this project, which focused on the elements that made up shapes, reviewed patterns and previously studied figures from the perspective of symmetry, and captured their properties. Based on the patterns created in the "Monkiri" game, the relationship between the number of folds and the characteristics of the shapes was investigated, and the students noticed the mathematical characteristics such as linear and point symmetry in even-numbered folds (Figure 5.3). In addition, in the unit "Our Future through Data," students earned about the development of science and technology in social studies, mathematics, and science as part of their cross-curricular learning to consider human ethics and the future of society. In the context of learning to formulate business plans to solve social problems developed in social studies, the aim was to critically review the data they had collected and their methods of analysis considering their objectives from various perspectives such as proportion, change, and distribution. These cross-curricular and authentic context-aware studies were assessed with respect to the development of

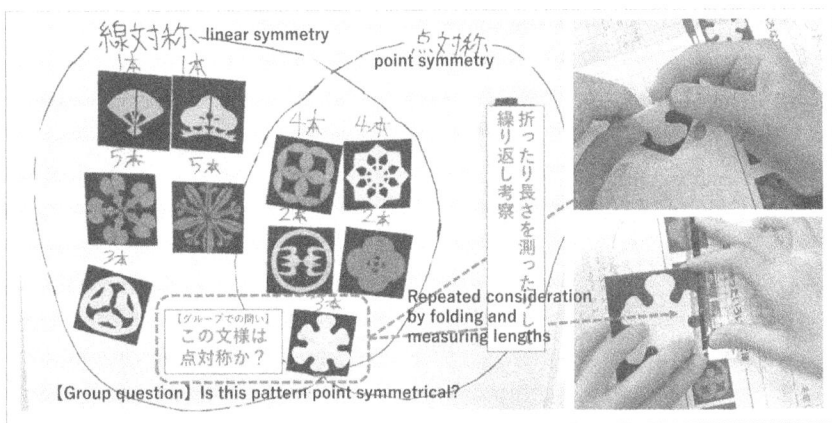

Figure 5.3 Whiteboard for group work.

Source: Fukuoka Elementary School attached to Fukuoka University of Education (2023, p. 53).

mathematical thinking that the mathematics department of the school wanted to foster over the long term across units. There was a general rubric for mathematics departments, as shown in Table 5.3, which was customized for each unit, and the degree of proficiency in thinking in each unit was assessed.

Table 5.3 Generic rubric for mathematics (creativity).

	Finding the question	*Logical thinking and mathematical expression*	
Stage 5	The student is able to find the correct mathematical elements of concrete situations, simplify and idealize them, make them more general, and refine the question.	The student is able to select a method of reasoning according to the situation, explain to a conclusion, develop the situation, discuss it, and confirm the validity of the result.	The student is able to express their own ideas and relate them to the ideas of others, using a variety of methods of expression according to the purpose.
Stage 4	The student is able to find, simplify, and idealize mathematical elements in concrete situations and correctly make mathematical questions.	The student is able to select a method of reasoning appropriate to the situation according to his/her own outlook, and is able to give a more general explanation while developing the scene.	The student is able to express their own ideas using methods of expression suited to the purpose, and compare their ideas with those of others and incorporate the merits of their own ideas.
Stage 3	The student is able to find the elements of mathematics in concrete situations, and make them into mathematical questions in his/her own way.	The student is able to choose a method of reasoning appropriate to the situation and develop the scene, but jump to conclusions in certain areas. (Some explanations are lacking.)	The student is able to express his/her own ideas and exchange opinions with others using methods of expression suited to the purpose.
Stage 2	The student is able to find all the mathematical elements necessary to solve a problem in a concrete situation.	The student is able to have a perspective, choose a method of reasoning appropriate to the situation, and draw conclusions.	The student is able to express his/her ideas using concrete objects and diagrams according to phenomena such as numbers, quantities, shapes, and their relationships.
Stage 1	The student is able to find some of the mathematical elements necessary to solve the problem.	The student is able to choose to reason about the situation with the help of the teacher and other learners.	The student is able to express his/her own ideas using specified concrete objects and diagrams.

Source: Materials provided by Fukuoka Elementary School attached to Fukuoka University of Education.

Although the Super Science High Schools (SSH) project and other programs have originally been characterized as STEM education, STEM/STEAM education is being developed as an initiative to nurture talented science and technology professionals who can tackle social problems, create value using cutting-edge technology, and who are aspiring creators of society, in active collaboration with the community, universities, and businesses outside the school, by actively utilizing the period for integrated study and project study in high schools. For example, the Hyogo Prefectural Board of Education (2023) is developing "Hyogo STEAM" in several high schools in the prefecture. The program emphasizes "independent exploratory activities based on knowledge of advanced technology obtained through collaboration with companies, etc., in collaboration with others to solve familiar problems and create new value." The program emphasizes the importance of studying advanced technology as a tool implemented in the real world, valuing objective data and evidence in finding and solving problems, developing not only theories and ideas but also models and plans and communicating these ideas to the outside world using English and other languages. At Kakogawa Higashi High School in Hyogo Prefecture, regardless of the acronym STEAM, a variety of special courses have been established in collaboration with universities, companies, NPOs, etc., on topics such as social implementation of advanced technology and regional design with "excitement" as the keyword. This was done to deepen the conventional project studies. Since various extracurricular programs and hands-on activities are offered, not only in the subject areas but also in the period for integrated study, the evaluation of the program is based on a complex data set that includes satisfaction questionnaires and free-text surveys for each course, a questionnaire survey that analytically questions students' sense of growth in the quality and ability they aim for, and tests aimed at measuring generic skills. The assessment was implemented in a program evaluation manner.

It is important not only to quantitatively and empirically capture the outcomes of combined efforts using data but also to qualitatively and narratively grasp and interpret the specifics of learning and growth in project studies, etc., in which student development across the curriculum is expressed intensively. In period for integrated studies and project studies, it is common for students to reflect on the process of inquiry-based learning over a long period, keeping a portfolio of their learning, and receiving coaching or tutorial guidance from teachers and others, also using rubrics that are aligned with the cycle of inquiry as a cue. For example, at Nara Women's Secondary School, which has long promoted advanced SSH while also emphasizing liberal arts, students move from exploring familiar themes in the 1st and 2nd grade to exploring themes from the perspective of Education for Sustainable Development (ESD) and Sustainable Development Goals (SDGs) in the 3rd grade. In 4th grade, they are given an introduction to project study where students begin to study inquiry methods in the natural sciences and humanities in earnest. Then, in 5th grade, students experience acquiring perspectives that transcend the boundaries between the humanities and sciences, and in 6th grade, they return to their fields of specialization to culminate their exploratory activities (Fujino & Hase, 2022). The "Project Study Roadmap" of Nara Women's University Secondary School (see Table 5.4 on pages 79–82), which organizes the quality and ability that the school's inquiry activities aim for, is rubric-like in nature, however, its primary focus is to show the process and prospects for qualitative improvement in inquiry, rather than simply showing the goals and criteria for learning through project study. It was created based on the specific assumptions of actual student learning at the school, and is used as a guide for teachers and students to connect and support the process of trial and error in inquiry as a story while imagining high-quality activities as a role model and the content of activities appropriate to each student's situation and developmental stage.

When it comes to the assessment of project studies, etc., it is often done by creating a kind of rubric of generic skills, and based on that table, quantitative and qualitative evidence is collected to analytically explain the student's learning situation. There, the student's learning and transformation remain analytically broken down, and no concrete picture of the student emerges. The generic skills, which are the components of quality and ability, are like radiographs, and their meaning and student's challenges become clear when they are interpreted in relation to the student in the flesh. In addition to an analytical and objective assessment, can we tell the overall specific story of learning that the project study produced, to the extent that we focus on the facts of a typical student, based on photos, videos, episodes of student learning, and reports and works as artifacts of learning? The point at which teachers can tell the story of growth in the form of student, and students themselves can tell their own stories of growth is the test of whether or not the curriculum development has substance and can compete based on students' learning and growth.

Conclusion

To question assessment is to ascertain whether there is solid learning and growth through educational activities. As mentioned at the beginning of this chapter, the question of assessment in STEM/STEAM education must be confronted with the question of its goals (learning outcomes) and educational value. In the magnetic field of Japanese-style competency-based curriculum reform, questions about learning outcomes and educational value are psychologized and abstracted, and this may contribute to the proliferation of curricula and initiatives in STEM/STEAM education. Furthermore, learning and growth are not only objectively understood numerically, but are ultimately examined in the concrete forms of students, with interpretations and value judgments about what constitutes learning and growth. How will STEM/STEAM education expose students to different worlds and knowledge (content), thereby broadening and enhancing the breadth and perspective of students' learning and lives? It is important to discuss specifically goals and assessment, which are often discussed in abstract and formal terms as theories of competence, in the layers of content organization and children's actual learning experiences.

Table 5.4 "Project Study Roadmap" of Nara Women's University Secondary School (prepared in 2018).

Approach to inquiry activities		Stage 1 *Learn the techniques of inquiry activities*	Stage 2 *Conduct inquiry activities with an emphasis on mathematical and scientific interpretation*	Stage 3 *Conduct advanced exploration activities that are not limited to the scope of study in high school*	Expert *Conduct original and progressive research activities based on your academic background*
(1) Setting of research topics [Problem finding].	Finding of research topics	The student can identify a research topic of interest that leads to the subject of inquiry activities.			The student is able to set a research topic of high social and academic significance.
	Examination of research topics	The student is able to select a verifiable research topic for problem setting.	In setting problem, the student can choose a research topic that allow for deeper mathematical and scientific interpretation.	The student can take on advanced challenges that are not limited to the scope of study in high school.	In setting problem, not limited to what he/she has learnt, the student can challenge a research topic that is beyond the level of high school students.
	Survey of previous studies	The student is able to research previous studies and find information necessary for inquiry activities.	The student is able to research previous studies and to set up problems that can be analyzed theoretically and experimentally based on what has already been learned.	The student is able to research previous studies and to set up problems that can be analyzed theoretically and experimentally based on the necessary knowledge.	The student is able to research previous research and find unresolved problems, and to set new problems from an original perspective.
	Appropriateness of issues	If the problem is too challenging, the student can re-set the problem at an appropriate level to match his/her exploration skills.	If the problem is challenging, the student can set the problem at the appropriate level while learning the necessary knowledge.	If the problem is challenging, the student can try the problem while learning the necessary knowledge.	If the problem is challenging, the student is able to acquire more advanced knowledge and skills in order to accomplish the task originally set.

(Continued)

Table 5.4 (Continued)

Approach to inquiry activities		Stage 1 — Learn the techniques of inquiry activities	Stage 2 — Conduct inquiry activities with an emphasis on mathematical and scientific interpretation	Stage 3 — Conduct advanced exploration activities that are not limited to the scope of study in high school	Expert — Conduct original and progressive research activities based on your academic background
(2) Research activities [The importance of "how" at each stage]	Methodology construction	The student can find the appropriate research method to solve the problem.	The student can use knowledge of mathematics and science to find analytical research methods.	The student can learn the necessary research methods while constructing more advanced research methods.	The student can construct new survey methods or add unique perspectives to existing methods.
		The student can develop an understanding of experimental equipment and theories that are being used for first time.	The student can select appropriate laboratory equipment and make logical interpretations.	In addition to selecting appropriate experimental equipment and logical interpretation, the student can challenge the construction of experimental apparatus and logic as needed.	The student can build their own experimental apparatuses suitable for their research activities and to construct their own logic for validation.
(3) Data processing and analysis [Collection of data and Information]	Data processing	The student can organize the obtained data in appropriate graphs and tables in order to read the trends shown by the data.	The student can organize the obtained data in appropriate graphs and tables in order to read the mathematical trends shown by the data.	The student can organize the obtained data in appropriate graphs and tables in order to read the mathematical trends shown by the data.	The student can perform mathematical operations necessary for data analysis, such as statistical significance...to perform logical analysis that goes beyond the knowledge of high school students.
	Analysis	The student can compare the results with those of previous studies and discuss sources of error.	The student can analytically consider the sources of error and approaches to improve them by comparing the results with those of previous studies.	The student can analytically consider the sources of error and approaches to improve them by comparing the results with those of previous studies.	Even for studies that have not been previously studied, the student can analyze error factors from a variety of perspectives and try more advanced approaches.

(Continued)

Table 5.4 (Continued)

Approach to inquiry activities		Stage 1 *Learn the techniques of inquiry activities*	Stage 2 *Conduct inquiry activities with an emphasis on mathematical and scientific interpretation*	Stage 3 *Conduct advanced exploration activities that are not limited to the scope of study in high school*	Expert *Conduct original and progressive research activities based on your academic background*
(4) Discussion and conclusion	Comparison with previous studies	The student can find results similar to previous studies.	The student can evaluate the validity of results based on previous studies with mathematical interpretation.	The student can evaluate the validity of results based on previous studies while learning the necessary mathematical interpretations.	The student can provide an original interpretation of the relevance of the research question with reference to multiple previous studies.
	Conclusion	The student can draw conclusions about the research question from the results obtained.	The student can find analytical conclusions from obtained results, including mathematical interpretations.	The student can find analytical conclusions from the results obtained, including interpretations beyond the knowledge of high school students.	The student can discuss the development of conclusions in depth from a scholarly point of view.
(5) Recording and presentation	Record of activities	The student is able to record activities in a research notebook.	The student is able to record activities in a research notebook in an organized manner so that others can understand them.	The student is able to record activities in an organized manner in a research notebook and provide analytical reflections for the next activity.	The student is be able to record activities using digital contents, not limited to research notebooks, and organize them for use in presentation activities.
	Presentation	The student is able to create a poster summarizing the results of the research.	The student is able to prepare a poster that analytically summarizes the results of the research. The student is able to create posters with an emphasis on discussion.	The student is able to create posters that summarize difficult research methods in an easy-to-understand manner. The student is able to summarize the results of research in the form of a thesis.	The student is able to prepare posters and papers at a level that can be presented at contests and conferences.

(Continued)

Table 5.4 (Continued)

Approach to inquiry activities		Stage 1	Stage 2	Stage 3	Expert
		Learn the techniques of inquiry activities	*Conduct inquiry activities with an emphasis on mathematical and scientific interpretation*	*Conduct advanced exploration activities that are not limited to the scope of study in high school*	*Conduct original and progressive research activities based on your academic background*
(6) Co-creation	Collaboration	The student can plan each research project and reflect on the content of the research in consultation with the faculty advisor.	The student can plan and reflect on the research project while collaborating with group members. The student can seek the advice of a faculty advisor at the appropriate time.	The student can plan and reflect on the research project on his/her own initiative in consultation with his/her advisor at appropriate times.	The student can exchange opinions not only with his/her advisors, but also with researchers, university faculty, and other experts as needed.
		The student is able to work within a group, dividing roles necessary, to carry out his/her research project.	The student is able to share roles and responsibilities within a group while utilizing each other's areas of expertise to advance his/her research project.	The student is able to improve his/her research project through discussions with students who have been involved in inquiry activities for long time, such as students in the science research group.	The student is able to discuss research topics with students involved in research in different fields to gain perspectives linked to multiple fields of study.
	Observation	The student is able to be interested in the methods of other people's inquiry activities and to identify similarities and dissimilarities with their own activities.	The student is able to observe the methods of other people's inquiry activities, consider how to apply the excellent methods of others to their own activities, and gain new perspectives.	The student is able to observe the methods of inquiry activities of students working on advanced tasks and students in science research groups to gain perspectives to enhance their own activities.	The student is able to observe the methods of inquiry of teachers, students involved in different fields of inquiry, researchers, etc., and establish research methods that the student do not know.

Source: Materials provided by Nara Women's University Secondary School.

References

Asano, D. (2022). Shakaihenyou to tankyu-modo eno chousen [Social transformation and the challenge of inquiry mode]. In M. Tamura & M. Sato (Eds.), *Tankyu-mode eno chousen [The challenge of inquiry mode]* (pp. 101–139). Ningendo [in Japanese].

Fujino, T., & Hase, K. (2022). Tayou na tashaga kyoumei suru tankyu katsudou [Inquiry activities that resonate with diverse others]. In T. Ishii (Ed.), *Koutougakkou shinsei no manabi, jugyou no fukami – jugyou no takumitachi ga teiansuru korekara no manabi [High school: Authentic learning, depth of teaching—Future teaching proposed by classroom experts]* (pp. 266–275). Gakujishuppan [in Japanese].

Fukumoto, Y. (2020). Kagaku to nijichou seikatsu no kakawari wo fukaku tankyu suru – tangen "Koubunshi kagoubutsu no seishitu to riyou" [Inquiring deeply into the relationship between chemistry and daily life—Unit "Properties and Uses of Polymer Compounds"]. In K. Nishioka (Ed.), *Koutougakkou Kyoka to tankyu no atarashii gakushu-hyouka [High school: New learning assessment of subjects and inquiries]* (pp. 102–107). Gakujishuppan [in Japanese].

Fukuoka Elementary School attached to Fukuoka University of Education. (2023). *Reiwa 4nendo kenkyu kaihatsu jissihoukokusho shiryouhen [2022 Research and development implementation report document]*. Fukuoka Elementary School Attached to Fukuoka University of Education [in Japanese].

Hyogo Prefectural Board of Education. (2023). *Hyogo ban STEAM kyouiku ni tsuite [Hyogo version of STEAM education]*. Document submitted to the 4th Working Group on the Future of High School Education [in Japanese].

Ishii, T. (2017). Theories based on models of academic achievement and competency. In K. Tanaka, K. Nishioka, & T. Ishii (Eds.), *Curriculum, Instruction, and assessment in Japan: Beyond lesson study*. Routledge.

Ishii, T. (2020). *Zouho-ban gendai america ni okeru gakuryokukeisei no tenkai-standard ni motoduku carriculam no settei [The development of academic achievement formation theory in contemporary America: Curriculum design based on standards (re-edited and expanded edition)]*. Toshindo [in Japanese].

Ishii, T. (2022). Competency base no kyouikukaikaku no kadaito tenbou-shokugyou kunren wo koete shakai eno ikou to otonatositeno jiritsu no tameno kyouiku e [Challenges and prospects of competency-based educational reform: Beyond vocational training to education for transition to society and independence as an adult]. *Japan Journal of Labor Studies, 742,* 16–27 [in Japanese].

Isobe, M., & Yamazaki, S. (2015). Design and Technology karano England STEM kyouiku no genjou to kadai [Current status and challenges of STEM education in England from design and technology]. *Journal of Science Education in Japan, 39*(2), 86–93 [in Japanese].

Isozaki, T., & Isozaki, T. (2021). Nihongata STEM kyouiku no koushiku ni mukete no rironteki kenkyu [Theoretical research for establishing a Japanese-style STEM education Analysis from a comparative historical point of view]. *Journal of Science Education in Japan, 45*(2), 142–154 [in Japanese].

Kumano, Y. (2014). Kagaku gijutsu governance to STEM kyouiku – nihon ni okeru governance-ron to america ni okeru aratana kagaku gijutsukaikaku karano kanten [Science and technology governance and STEM education: Perspectives from governance theory in Japan and new science education reforms in the U.S.] In *Kagaku gijutsu governance no keisei no tameno kagaku kyoiku-ron no kouchiku ni kansuru kisotekikenkyu [Basic research on the construction of science education theory for the formation of science and technology governance final report]* (pp. 1–16) [in Japanese].

Marzano, R. J. (1992). *A different kind of classroom: Teaching with dimensions of learning.* Association for Supervision and Curriculum Development.

Matsubara, K. (2020). Shishitsu nouryoku no ikusei wo jushi suru kyouka oudanteki na manabi to STEM/STEAM kyouiku [Cross-curricular learning and STEM/STEAM education emphasize the development of qualities and abilities]. *Nihon kagakukyouiku gakkai dai44kai nenkanronbunshu [Proceedings of the 44th annual meeting of the Japanese society for science education]* (pp. 9–12) [in Japanese].

Matsubara, K., & Kosaka, M. (2017). Shishitsu nouryoku no ikusei wo jushi suru kyoukaoudantekina gakushu tositeno STEM kyouiku to toi [STEM education and questions as cross-curricular learning emphasize the development of qualities and abilities]. *Journal of Science Education in Japan, 41*(2), 150–160 [in Japanese].

Matsushita, K. (Ed.). (2010). *"Atarashi nouryoku" wa kyouiku wo kaerunoka [Will "New Competencies" change education]?* Minerva Shobo [in Japanese].

Ministry of Economy, Trade and Industry (METI). (2019). *'Future Classrooms' and EdTech Study Group Second Proposal* [in Japanese]. https://www.meti.go.jp/shingikai/mono_info_service/mirai_kyoshitsu/pdf/20190625_report.pdf

Nishioka, K. (2016). *Kyouka to sougougakushu no curriculam sekkei – performance hyouka wo dou ikasuka [Curriculum design for subject and integrated learning: How to utilize performance assessment].* Toshobunka [in Japanese].

Oshima, R. (2022). Singapore. In T, Suzuki (Ed.), *Gakkou ni okeru kyouikukatei-hensei no jisshouteki kenkyu houkokusho 4 shogaikokuno senshintekina kagakukyouiku ni kansuru kisoteki kenkyu [An empirical study of curriculum development in schools report 4: A basic study of advanced science education in other countries: Focusing on scientific inquiry and STEM/STEAM].* National Institute for Educational Policy Research [in Japanese].

Rychen, D. S., & Salganik, L. H. (Eds). (2003). *Key competencies for a successful life and a well-functioning society.* Hogrefe & Huber.

Suzuki, T. (2022). *Gakkou ni okeru kyouikukatei hensei no jisshouteki kenkyu houkokusho4 shogaikoku no senshintekina kagakukyouiku ni kansuru kisoteki kenkyu-Kagakuteki tankyu to STEM/STEAM kyouiku wo chushin ni [An empirical study of curriculum development in schools report 4: Basic research on advanced science education in other countries: Focusing on scientific inquiry and STEM/STEAM].* National Institute for Educational Policy Research [in Japanese].

Vasquez, J., Sneider, C., & Comer, M. (2013). *STEM lesson essentials, grades 3–8: Integrating science, technology, engineering, and mathematics.* Heinemann.

Wiggins, G., & McTighe, J. (2005). *Understanding by design expanded* (2nd ed.). Association for Supervision and Curriculum Development.

6 Tasks that drive students to learning processes toward knowledge integration

Moegi Saito

Introduction

This chapter demonstrates findings about tasks that effectively support the students' learning process toward knowledge integration.

Recently, STEM education has been attracting attention as a perspective from which to rethink education in the fields of science, technology, mathematics, and engineering. One of the key ideas that STEM education offers is the idea of considering "Knowledge Integration (KI)" as a core of learning that we wish to create in classrooms. The KI here is not necessarily limited to knowledge integration across subject areas (e.g., linking math and science). For example, in the subject of science, it is important to connect knowledge of various qualities and experience, both of which students are exposed to in the classroom, such as connecting "real-world problem solving" and "scientific knowledge" (Saito, 2020). KI across subject areas is gradually being promoted based on intra-subject KI. Therefore, to understand the long-term learning process of STEM, it is important to understand the process of KI within one domain. Similarly, the field of learning sciences, which studies the processes and mechanisms of human learning, has shown that deep understanding of the learning content is supported by the relation-making among knowledge and integration of knowledge by the learners themselves (Clark & Linn, 2013). Supporting the process of learning toward KI in the classroom is an important issue for the future of science instruction.

On the other hand, it is difficult to identify the causal relationship between a teaching strategy and the learning outcomes, regardless of the nature of the tasks and their settings. Various factors, both intentional and unintentional, influence the learning outcomes achieved in the classroom, such as students' previous knowledge, thinking ability, and familiarity with the teaching method. Therefore, in this chapter, I rely on research findings from cognitive science. Cognitive science has already identified what kinds of thinking and dialogue processes lead to KI. Therefore, I first analyze the dialogues in the lesson where, judging from the students' writings at the end of the lesson, KI is deemed to have been achieved, and confirm whether ideas and dialogues in the direction of KI were generated during the lesson. Then, I consider the characteristics of the task used in that class. Through these two steps, my aim

DOI: 10.4324/9781003392545-7

is to gain more elaborated findings than before on task-setting that effectively supports the learners' learning process toward KI.

Patterns of thought and dialogue that tend to lead to KI—from previous studies in cognitive sciences

This study focuses on "collaborative inquiry" into questions in the process of problem-solving as one of the most promising candidates for patterns of thinking and dialogue that are likely to lead to KI.

Cognitive science has accumulated a large amount of knowledge about the characteristics of dialogue that contribute to the deep understanding of the participants in order to elucidate the process and mechanism of human learning. For example, Barnes (1976) contrasted "exploratory talk" and "presentational talk," and found that the former type of talk tended to promote deeper understanding. "Exploratory talk" is dialogue that includes repetition questions such as "What?" or "If that's the case, how about this?" and is said to often produce comparative examination and consideration of ideas. The concept of "inquiry-based dialogue," which was proposed by Barnes, was later extended to dialogue that encourages knowledge sharing and review through the free exchange of ideas at a stage in the collaborative problem-solving process when the answer is not clear.

Tohyama and Shirouzu (2017) also organized such studies, and drawing on the current consensus of cognitive science research, we can see that a shared view exists that patterning dialogue, in which incomplete, question-generating utterances are frequently exchanged between speakers, is more likely to lead to deeper understanding, as opposed to non-involving patterning dialogue such as the exchange of opinions, full criticism, and non-critical acceptance. In fact, Chi and Menekse (2015), in their analysis of the relationship between the degree of deepening of conceptual understanding in pre- and post-task-solving tests and dialogue patterns, confirmed that "co-constructive dialogue" (dialogue in which one member digs deeper into what the other member had said before) is associated with deeper understanding.

From these findings, we can find a common view that pattern dialogues with frequent speaker exchanges of incomplete, question-generating utterances are more likely to lead to deeper understanding. Dialogues referred to as "collaborative inquiry" in which questions are spontaneously raised among learners who work together to solve problems, solutions are proposed to those questions, and further questions are generated, are considered to be beneficial for learners' knowledge integration.

This chapter presents a case study of two science classes. We aim to verify through the case study, whether "collaborative inquiry into questions" actually occurs frequently in classes where KI is achieved through collaborative problem-solving activities. Specifically, and independently of the purpose of this chapter, using the in-class dialogue transcripts of two cooperative problem-solving science classes conducted by the same teacher in the same school using the same methodology, we show that the class with higher achievement in KI had a higher percentage of questions spontaneously generated by the learners and of collaborative inquiry by the group's members during the problem-solving.

Then, by examining what features of the design and support of the target class supported the "collaborative inquiry," we would like to obtain suggestions for triggering learning that moves toward KI. To this end, we will examine the impact of lesson design on the generation of collaborative inquiry by contrasting the characteristics of lesson design and dialogue, focusing on the subtle differences between the two classes.

The targets of analysis

Knowledge Constructive Jigsaw—"The Mechanics of Exercise"

The lesson design method used in the two cases was the "Knowledge Constructive Jigsaw" ("KCJ"). The KCJ is a method that promotes deeper understanding among diverse participants through the flow of cooperative problem-solving in five steps, and its effectiveness has been confirmed in a variety of subjects (Miyake, 2013).

The KCJ consists of five learning activities: (1) writing an answer to the day's given problem; (2) an expert-group activity which allows each individual learner to accumulate pieces of knowledge relevant in solving the problem; (3) a jigsaw-type activity where learners from different expert groups get together to exchange and integrate the pieces of knowledge and form an answer; (4) a cross-talk activity where the learners exchange their ideas for solutions, involving the entire class; and (5) writing down an individual answer again to the same problem. Like the original Jigsaw method (Aronson, 1978), this method distributes knowledge among the learning partners in Step 2 for integration in Step 3; however, it places greater emphasis on the role of a shared "problem" for knowledge construction by adding Steps 1 and 4, as well as that of the changes in thinking of each individual by adding Steps 1 and 5. The most important activity in KCJ is the jigsaw-type activity (Step 3), which fosters constructive interaction to deepen understanding based on the individual experiences of Steps 1 and 2.

Next, a summary of the lesson design for "The Mechanics of Exercise" (2014 and 2016) is shown in Table 6.1. The lessons were designed for 8th graders' science classes. The goal of the lesson was to understand the mechanics of body movement which included three points: the sensory nervous system, commands from the brain, and the motor function of the nervous system. The expert activities consisted of reading materials and diagrams and answering short questions on the following three topics: (A) the nervous system; (B) the skeletal structure; and (C) the mechanism of muscles. The printouts that were distributed were identical for the two lessons.

On the other hand, the main presented task differed significantly between 2014 and 2016. The 2014 assignment was an explanation in response to the question, "What do you have to do to catch a falling ruler?"

Table 6.1 KCJ "The Mechanics of Exercise" (2014 and 2016).

Main task	2014: Explain the "body movements of a tennis player hitting a tennis ball." 2016: Give an explanation in response to the question, "What do you have to do to catch a falling ruler?"
Expert materials A	Nervous system: • Sensory organ → sensory nerve → spinal cord → brain: information flow • Brain → spinal cord → motor nerves → muscle: chain of command
Expert materials B	Skeletal structure: • Skeletal structure (endoskeleton) and joints
Expert materials C	Mechanism of muscles: • How muscles are built and work • Muscle and bone attachments (tendons)
The goal	The expected answer should include the three points below: • A description using diagrams and words of the stimuli received, the organs involved, and the route they take to the brain (knowledge of the sensory nervous system) • Expressing in words what the brain thinks and what it commands (knowledge of what the brain commands) • Description of the route from the brain to the muscles and the movement of muscles and joints using words and diagrams (knowledge of the motor nervous system)

In 2014, although the subject matter involved complex movements involving multiple nerves, the skeleton and muscles, the task was designed to easily capture the students' interest and evoke familiar experiences for them to think about. By contrast, in 2016, a classification was used that divided the number of solutions to be analyzed, and a task was set up with subject matter that could be visually reproduced many times.

Lesson outcomes

Next, we show the differences in students' deepening of understanding in both lessons. Figure 6.1 compares the extent to which students referred to knowledge in their explanations of the task, based on their answers pre- and post-lesson.

The answers were analyzed for 12 students in the 2014 lesson and 38 students in the 2016 lesson. In the analysis, we divided the students' answers into either "complete" or "partial." The "complete" answer about the nerves gave an answer that described the route from the sensory organ that receives the stimulus to the cerebrum without excess or deficiency. The "complete" answer about the brain gave an answer that correctly described both the content of the stimulus received (the ruler started to move) and the content of the command (the muscles of the index finger and thumb contracted). The "complete" answer about commands gave an answer that described the path from the brain's commands to the muscles that move without excess or

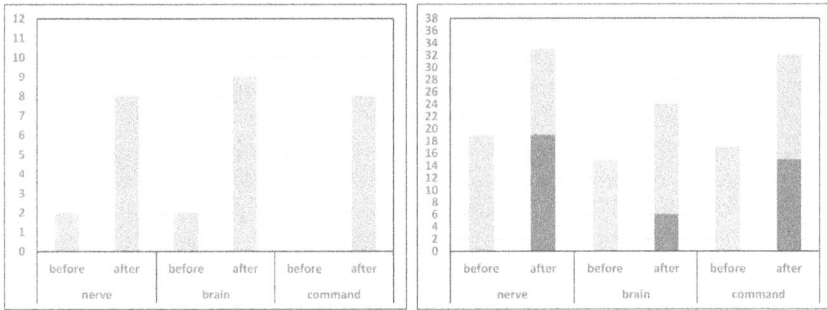

Figure 6.1 Changes in understanding before and after the lesson (left: 2014, right: 2016) $N = 12$ in 2014 and $N = 38$ in 2016; black indicates complete answers and gray indicates incomplete answers (black: complete, gray: partial).

deficiency. Answers that included omissions or errors in any of these three elements were marked as "incomplete" even if they referred to relevant knowledge.

From the graph, it can be seen that the number of students who were able to give an explanation that integrated the three pieces of knowledge into the task in the 2016 lesson increased significantly after the lesson. In addition, before classes, seven students were able to integrate all three pieces of knowledge into the task, while 22 students were able to do so after the classes. On the other hand, in the 2014 classes, the lesson was not successful. According to an analysis of the students' post-lesson results, none of them were able to write "complete" answers which included these three points. The teacher reviewed the lesson, redesigned a second version of the lesson and taught it in 2016. In the second time round, the results of the students improved. Even taking into account the difference in the number of analyzed answers, it can be said that the 2016 lesson tended to have relatively higher results in KI as indicated by the feedback of the practitioners.

Now, let us take a closer look at the specifics of KI in the 2016 classes. Table 6.2 shows examples of the pre- and post-lesson answers of three students who worked on the task in the same group in a jigsaw activity.

All three students wrote insufficient answers before the lesson when compared to the teacher's expectations. On the other hand, after the lesson, the students were able to explain the mechanism of a series of exercises by integrating the three pieces of knowledge in line with the real-world task of "catching a falling ruler with their fingers." For example, although Student X referred to knowledge of the "sensory nervous system," "brain," and "somatic nervous system" before the class, there were incomplete pathways in both the sensory and somatic nervous systems, and the content of the brain's commands was also not described. After the lesson, they were able to connect the scientific term "stimulus" with the phenomenon of the "falling ruler" and explain in more detail the process of transmission of the "stimulus" by using such related terms as "sensory organs" and "sensory nerves." We are

Table 6.2 Examples of pre- and post-lesson answers of "The Mechanics of Exercise" (2016).

Pre	Post
X	Eye→Brain→Hand Vision Command (light)	Central nervous system ···it sends a signal to the brain. Stimulus ⟶ Sensory Organs ⟶ Sensory Nervous ⟶ Sensory Nervous ⟶ spinal cord ⟶ brain (Video of ruler falling) (Eyes) (peripheral nervous) (peripheral nervous) Response to the stimulus ⟵ The muscles ⟵ Motor Nervous It analyzes signals sent from the brain and sends the correct commands to the muscles.
Y	The five senses sense that the ruler is falling, they send what they sense to the brain, and the brain sends instructions to the fingers to "move and grasp the ruler."	Eye → sensory nerve → calculus ⇔ brain (received the signal that the ruler fell and gave the order to the finger muscles to grab the ruler) ↓ Motor nerves → Hand muscles
Z	The muscles of the fingers stretch and contract, and the joints move. Commands from the brain are issued to the nerves in the fingers.	The eye receives a light stimulus → it sends a signal to the brain that "the ruler is falling!" → the brain decides, "Grab the ruler!" → the brain sends the command to the muscles, etc. necessary to grasp the ruler.

also able to infer that Student Y and Student Z were not very aware of the connections between the task and knowledge, and between different kinds of knowledge, before the lesson, from the fact that no conjunctions or arrows expressing connections were included. By contrast, after the lesson, they were able to explain the connections between events in the process from stimulus to response. Student Z's explanation of the "sensory nervous system" and "somatic nervous system" pathways remained incomplete. Nevertheless, it can be said that the degree of KI had increased significantly for all three students throughout the course of the lesson.

The learning process toward KI

Data analysis

Data

Comparing the results of the 2014 and 2016 classes, it was observed that knowledge integration was more advanced in 2016. What, then, was the "learning process toward knowledge integration" that occurred in the 2016 classes? We would like to make a comparison by analyzing the dialogue of the "jigsaw activity" (Step 3) in the respective classes of 2014 and 2016.

The data to be analyzed comprises the dialogues during the jigsaw activities for one class each in 2014 and 2016. The jigsaw-type activity (Step 3) was chosen for the analysis because it was thought to be the most likely to show the effects of problem-setting.

Group dialogues were recorded with an IC recorder placed in the center of each group. Eight groups were recorded in 2014 and six groups in 2016. Six groups with clear recordings were selected for each year with a total of 12 groups being used in the analysis. Each jigsaw activity lasted 15 minutes. All of the speech data was transcribed manually, differentiating the speakers as much as possible, and the utterances were converted to text for each speaker. The utterance uttered in one breath was categorized as "one utterance." A total of 3,166 utterances from the six groups in 2014 and 4,195 utterances from the six groups in 2016 were used for the analysis.

Procedure for the analysis

The dialogue analysis was conducted in two steps: identification of the "questions" in the dialogue, and coding of the utterances following the "questions."

a   Step 1. Identification of the "questions" in the dialogue

First, the content and flow of each group's dialogues were confirmed, and the utterances that could be identified as spontaneous "questions" from the group members were identified. The analysis was conducted by two researchers, including the author of this chapter, who identified "questions" as utterances that could be questions when converted into text, such as questions, queries, inquiries, and confirmations. The two researchers analyzed a total of 12 groups in 2014 and 2016, sharing the analysis of 339 utterances in 2014 and 430 utterances in 2016, which were identified as "questions." The two researchers consulted on the dialogues that needed to be reviewed in order to make a judgment.

b Step 2. Coding of utterances following the "question"

Next, we coded the utterances that followed the "question" as manifestations of the learning activity that followed the "question", and determined whether the spontaneous "question" was explored by the other group members, or not explored, whether it was explored collaboratively, responded to and completed, or explored by itself. Coding was done in "general categories" and "subcategories."

"General categories" are categories for determining whether a "question" is "shared" or not, and focusing on whether the other members of the group responded to the "question" in some way. If there was an utterance after the "question" that could be confirmed to be an utterance in response to the "question," such as the suggestion of a solution, restatement of a question, or return of a question, the utterance was considered "shared". If only irrelevant utterances other than those in response to the "question" were found after the "question," the utterance was considered "non-shared."

The "subcategories" are categories that focus on the content of the responses identified in the "general category" to determine if the "shared" questions were explored collaboratively. Cases in which a "question" was followed by a response from another member and no further statements or actions to explore a solution to the "question" could be confirmed were categorized as "ended with a response." Cases in which a "question" was not completed after a single response and multiple members were confirmed to have made statements or taken actions to explore the solution after the question was submitted were categorized as "collaborative inquiry." These members may or may not have included the person submitting the question.

Although not relevant to the analysis of "the percentage of questions spontaneously generated by students and collaboratively pursued as an inquiry by the group members during the problem solving," we also analyzed the subsequent activities when the spontaneous questions were not shared during the dialogue by creating "subcategories" of "self -inquiry" and "no inquiry" for the "non-shared" questions. The definitions of each category are shown in Table 6.3, and examples of these categories are shown in Tables 6.4 and 6.5.

Table 6.4 shows an example of a dialogue in which the "question" was shared after a member of the group had asked it. Speaker C says, "If it only had one joint, it would only move like this, but since it has several joints, does that mean it can bend?" and on receipt of this question, Speakers B and A proposed solutions such as, "Yes, yes, it can make delicate movements" and "It can actually move this far."

Table 6.3 Categories of dialogue analysis.

General category	Subcategory	Definition
Share	Ended with a response	After a question was submitted, other members responded once, and then no statements or actions to seek a solution to the question could be confirmed.
	Collaborative inquiry	After submitting a question, multiple members, including the submitter of the question, could be seen to be making statements and taking actions seeking for a solution.
Non-shared	Self-inquiry	After submitting a question, the submitter of the question could be confirmed to be making statements and taking actions seeking for a solution.
	No search	After the submission of the question, no one could be confirmed to be making statements or taking actions seeking for a solution.
Unable to categorize	—	—

When a question submitted by one of the members generated a statement that involved other members in proposing and exploring solutions, it was coded as a "collaborative inquiry." By contrast, Speaker C's question, "Are you finished?" was answered by Speaker A with "Yes," and no one continued

Table 6.4 Coding example of an activity created after a "question" (in the case of "sharing").

	Speaker A	Speaker B	Speaker C	General category	Subcategory
1			If you only do one.		
2	So far, in fact.		It only works like this.		
3		Yes, yes, yes, yes.	It has joints.		
4			You mean bent, right?	*Shared*	*Collaborative inquiry*
5		Yes, yes.			
6	It works so far in practice.	Able to make delicate movements.			
7			It means that it can be moved like this ...		
	Omission (of the middle part of the text)				
8			The end?	*Shared*	*Ended with a response*
9		Yeah ...			

Table 6.5 Coding example of an activity created after a "question" (in the case of "non-shared").

	Speaker A	Speaker B	General category	Subcategory	Speaker C
1					Just a sentence.
2		And if you want to paint, why don't you paint?	*Non-shared*	*No search*	
3					What?
		Omission (of the middle part of the text)			
4		What happened to your brain?	*Non-shared*	*Self-inquiry*	
5		Oh, your brain is totally opposite!			

with a related utterance. Such cases were coded as "ended with a response." By contrast, Table 6.5 shows an example of a dialogue in which the "question" was not shared after a member of the group had asked it.

Results

Table 6.6 shows the results of the dialogue analysis. As a result of the analysis, we found that in the 2016 class, which had higher learning outcomes, the percentage of collaborative inquiry during the task-solving was higher than in the 2014 class. In 2014, only 16% of the "questions" spontaneously posed during the Jigsaw activities led to joint exploration, whereas in 2016, 39% of the "questions" were used for spontaneous collaborative inquiry. This result shows that even in the case of junior high school students who are still in the process of development, in the process of learning toward KI, a question posed by one of the members invites the other members to propose or explore a solution, or

Table 6.6 Group members' reactions to questions spontaneously generated by the students in 2014 (*N* = 339) and 2016 (*N* = 430).

General categories	2014		2016		Subcategories	2014		2016	
	No.	%	No.	%		No.	%	No.	%
Shared	148	44%	295	69%	Ended with a response	95	28%	128	30%
					Collaborative inquiry	53	16%	167	39%
Non-shared	115	34%	67	16%	Self-inquiry	47	14%	39	9%
					Non-inquiry	68	20%	28	7%
Unable to categorize	76	22%	68	16%	—	0	0%	0	0%

N: Total number of questions spontaneously asked by members during the jigsaw activities.

even incomplete and question-generating dialogues appear frequently. It can be said that these results support the findings of previous studies.

Task settings to support the learning processes of students toward KI

Why, then, was the process of learning toward KI more likely to occur in the 2016 class than in the 2014 class? In this chapter, I focus on how the task was set, which was the biggest difference between the lesson in 2014 and 2016, and consider the reason for why the way in which the task was set in the 2016 lesson was more likely to provoke "collaborative inquiry", by referring again to previous research in cognitive science.

The main task of the 2014 lesson was to explain the "body movements of a tennis player hitting a tennis ball." Although the task was designed to engage the interest of the students and evoke familiar experiences, the subject matter involved complex movements involving multiple nerves, the skeleton and muscles. Furthermore, it was difficult for the students to intentionally reproduce the movements in the classroom. On the other hand, in the 2016 lesson, the students were assigned the task of giving an explanation in response to the question, "What do you have to do to catch a falling ruler?" The subject matter here was a simple movement that uses local nerves, bones, and muscles, and students could reproduce this over and over in the classroom in front of their own eyes. Comparing the two, it can be said that the characteristics of the task in 2016 were that it was easy to simplify the task in line with the three expected outcomes of knowledge (Table 6.1) that were the goals of the lesson, it was easy to think about the task while trying out the movements, and it was easy to encourage the expression and sharing of ideas.

Through previous research in cognitive science, it is known that there is a need to set a task, which is relatively simple, in order to be able to grasp what one is trying to understand at the starting point of understanding (Gelman, 2003). In addition, in collaborative problem-solving situations where multiple people are attempting to solve a problem together, research shows that setting a task where it is easy for a number of people to collectively, not individually, find a solution, encourages people to express and share their ideas, and stimulates dialogue that leads to a deeper understanding (Shirouzu et al., 2002). The way in which the task was set in the 2016 lesson fulfilled these two conditions sufficiently to provoke a dialogue that could be called "collaborative inquiry." Therefore, in order to effectively support the learning process toward KI, I hypothesize that it is useful to set simple and highly reproducible tasks.

As a starting point for hypothesis testing, let us take another look at specific examples of dialogue from 2014 and 2016. Table 6.7 compares the dialogue in the same scene in each class, a scene in which the task is addressed in a "jigsaw activity (Step 3)." The letters indicate the speaker, and the notes in parentheses are the author's notes.

Table 6.7 Specific examples of dialogue during the jigsaw activity.

2014 : Explaining the "body movements of a tennis player hitting a tennis ball"	2016 : Explaining in response to the question, "What do you have to do to catch a falling ruler?"
c : Then what, uh, feet and …	B : What happens before my eyes? Actually, not my eyes,
a : Because…	
b : Foot, hand, waist.	A : Huh? What was that?
c : Arms? You're talking about muscles now, right?	C : Stimulus, write about stimulus on our board.
a : Which muscle?	A : Iris?
b : All. I have lots of muscles all over my body.	B : Yes, "the stimulus" here, is light.
a : Which muscle …	C : "The stimulus" here is the light, and after the light is reflected, I see this thing.
b : Wrist, foot …	B : Light, light.
a : Arm.	C : I don't think you need to write that much.
b : I thought it was about the waist …?	A : It's complicated.
a : Then to the brain, to the brain?	C : From the lens of the eye. Was it the iris?
c : Move it. Hmmm.	B : Huh? But the stimulus is light. Stimuli go to the sensory organs. (Writing on whiteboard.)
a : Wrist, wrist, foot, waist no …	
c : Uh … feet, what feet?	C : Oh, you mean it's transmitted?
	B : Yes *(said in English)*.
	A : You don't have to respond in English.
	B : It's not the retina, it's the eye. Here (points to whiteboard)
	C : Since it is happening inside the body, isn't the light telling the brain what is happening?
	B : But…
	A : But it is visible by light.

In the jigsaw activity in 2016, the dialogue started from the question introduced by Student B, "What happened before my eyes? Actually not my eyes." Then Student A and Student C shared B's question and jointly participated in the inquiry. Through such "collaborative inquiry" it became clear that Student B was wondering what stimulates the eyes, and the connection of knowledge such as "stimulus" = "light" and "stimulus" is transmitted to "sensory organs" was found. "What happened before my eyes? Actually not my eyes." is a question that at first glance, one might not even know what the student is talking about. In spite of this, Students A and C did not ignore or complete the response with "I don't know", but shared B's question and participated in the inquiry collaboratively. The shared concrete image of "catching a falling ruler with one's fingertips" helped them retain the perspective that "we are talking about that experiment, so we should be able to figure it out if we all think about it together."

By contrast, in 2014, although the "brain" and "muscle" were the topics of discussion, each of the three participants thought of the head, hands, feet, and hips differently, leading Student C to express discomfort that the task was not being shared, saying, "We are talking about muscles now, aren't we?" The dialogue seemed to show difficulty in focusing on the connection of knowledge such as "brain," "instructions", and "muscle". This may be due to the fact that the task of "hitting a ball" is a complex exercise involving a great number of muscular and skeletal movements (compared to catching a ruler).

Based on a detailed examination of the dialogues, I propose that it would be worth further examining the usefulness of "setting simple and highly reproducible tasks."

Conclusion

In this chapter, I focused on the pattern of dialogue (e.g., "collaborative inquiry") that is characteristic of the learning process in which learners move toward KI in science classes, and discussed the requirements for setting a task that supports "collaborative inquiry." As a result, I was able to formulate a promising hypothesis that "setting simple and highly reproducible tasks is useful" to support the learning process toward KI.

In this chapter, I selected a science lesson, but this hypothesis could be applied to other subjects. For example, Shirouzu (2013) analyzed the learning processes of 6th-graders in a collaborative math class which aimed to integrate various expressions of fractions. In the class, the task of "Cut out 3/4 of 2/3 of a piece of origami paper" was set, and children discussed the task while looking at the origami paper they had actually cut. The task of cutting origami to the indicated size is simple and highly reproducible for learning fractions. Through the lesson, children jointly explored whether their individual ideas were the same or different, and came to understand that the area of each object was 1/2. In his analysis, Shirouzu (2013) pointed out that learners easily integrated a member's idea with their own when the member's "verbalizations were relevant to the task in hand." According to this point, the setting of a simple and highly reproducible task is considered to be likely to generate "collaborative inquiry" regardless of the subject by restricting the focus of individual learners within a certain scope.

Comparing the ways in which the task was posed in the classes analyzed in this chapter, it is thought that a large number of teachers could easily understand the merits of the set task in the 2014 lesson that connected with the motivation and interest of the students and familiar experiences. On the other hand, the effectiveness of the simple and highly reproducible task set in 2016 only becomes apparent when we focus on the learning process. Therefore, it is thought that many of the teachers did not notice the merits of this approach. In this chapter, through analyzing the dialogues, I have been able to show the possibilities of a task set in a way that may not seem remarkable from the teacher's perspective. In order to achieve new educational goals that are important from the perspective of STEM, we would need to rethink how the task is set by focusing not only on learning outcomes but also on the thinking and dialogue processes of the students.

References

Aronson, E. (1978). *The jigsaw classroom*. SAGE Publications.

Barnes, D. (1976). *From communication to curriculum*. Penguin.

Chi, M. T. H., & Menekse, M. (2015). Dialogue patterns in peer collaboration that promote learning. In L. Resnick, C. Asterhan, & S, Clarke (Eds.), *Socializing intelligence through academic talk and dialogue* (pp. 263–274). American Educational Research Association.

Clark, D. B. & Linn, M. C. (2013). The knowledge integration perspective connections across research and education. In S. Vosniadou (Ed.), *International handbook of research on conceptual change* (2nd ed., pp. 520–538). Taylor & Francis Group.

Gelman, S.A. (2003). *The essential child: Origins of essentialism in everyday thought*. Oxford University Press.

Miyake, N. (2013). Case report 5: Knowledge construction with technology in Japanese classrooms (CoREF). In P. Kampylis, N. Law, & Y. Punie (Eds.), *ICT-enabled innovation for learning in Europe and Asia* (pp. 78–90). European Commission, Joint Research Centre.

Saito, T. (2020). STEM/STEAM kyoiku no kouseigainen [Re-entry the Construct of STEM/STEAM Education]. *Japan Journal of Educational Technology*, *44*(3), 281–296 [in Japanese].

Shirouzu, H., Miyake, N., & Masukawa, H. (2002). Cognitively active externalization for situated reflection. *Cognitive Science*, *26*(4), 469–501.

Shirouzu, H. (2013). Focus-based constructive interaction. In D. D. Suthers, K. Lund, C. P. Rose, C. Teplovs, & N. Law (Eds.), *Productive multivocality in the analysis of group interactions (computer-supported collaborative learning series 16)* (pp. 103–122). Springer.

Tohyama, S. & Shirouzu, H. (2017). Kyocho mondai kaiketsu noryoku wo ikani hyoka suruka [Proposal for an assessment framework for collaborative problem solving skills: Cross-sectional analysis of dialogue data]. *Cognitive Studies*, *24*(4), 494–517 [in Japanese].

7 Food and nutrition education (*Shokuiku*) as a part of STEAM education

Takako Isozaki and Tetsuo Isozaki

Introduction

For the last few decades, there has been a global emphasis on food and nutrition education in response to a rise in health consciousness and the global imbalance between food supply and demand. Recognizing the significance of this trend, the World Health Organization (World Health Organization [WHO], 2004) released a diet and health strategy to promote and protect well-being through healthy eating and physical activity. The United States also launched the "5 A Day" campaign in 1991. This trend has also been observed in Japan, where the need and demand for food and nutrition education in schools have, consequently, increased.

The term "food literacy" is linked to food and nutrition education, which is known as *shokuiku* in Japan, and is occasionally used in the context of school education. According to Truman et al. (2017), studies on food literacy emerged around the beginning of the millennium when *shokuiku* was promoted as part of the Japanese government's campaigns to address widespread social issues, such as unbalanced nutrition and the global rise in obesity and lifestyle-related diseases.

Food literacy can be fostered in schools, which raises an important question: How should food and nutrition education be provided to develop food literacy in schools in international contexts? Stitt (1996) investigated food education in nine countries and argued that "*all* pupils should receive *comprehensive* food education... at school" (p. 34, italics in the original). McCloat and Caraher (2020), who analyzed international comparison data on food education in schools, suggested that food education should be provided by specialist teachers in an integrated and sequential manner through subjects such as home economics. Like education, food is based on a country's traditions and overall culture, creating a specific "food culture." Accordingly, food and nutrition education should be based on each country's educational and cultural contexts.

DOI: 10.4324/9781003392545-8

Research aim, questions, and methodology

This study investigated the methods of organizing *shokuiku* in Japanese schools from a historical perspective. To accomplish this aim, we formulated the following research questions: (1) What is *shokuiku?* and (2) How has *shokuiku* been taught in schools and, specifically, in home economics? In Japan and other countries, home economics is regarded as the main subject for teaching food-related knowledge, skills, and attitudes (McCloat & Caraher, 2016, 2020; Pendergast, 2012). We focused primarily on home economics given that it is considered the "linchpin" (McCloat & Caraher, 2016, p. 108) of a comprehensive education system. Knowledge of nutrition, one of the components of food literacy, is related to science; thus, we referred to the relationship between home economics and science where appropriate. Then, we analyzed historical data from the perspective of curriculum politics (Lawton, 1980) to determine a new organizational method for *shokuiku* in Japanese schools.

Theoretical framework: what are *Shokuiku* and food literacy in Japan?

What is shokuiku?

The Japanese word *shokuiku*, which can be traced back to the end of the 19th century, is translated as "food and nutrition education." In 1896, Sagen Ishizuka wrote the book *Kagakuteki shokuyō chōjyu-ron* (*A Chemical/Scientific Theory of Nutrition for Health and Longevity*), in which he underlined the fundamental significance of physical, intellectual, and talent education (which develops each child's individuality and talents) and the Japanese "traditional diet, cooking methods, and chemical/scientific [balanced] diet" (p. 276). Ishizuka's discourse on *shokuiku* focused primarily on home education rather than school education. This is because the attendance rate of school-age children in compulsory education (elementary school) in 1896 was only 64.22% (boys: 79.00%, girls: 47.53%) (Ministry of Education, Science and Culture [MoE], 1980), and there were no subjects related to food and nutrition education in the curriculum.

According to Morita (2004), the term *shokuiku* was not publicly used until the 1980s, and there was no concrete definition of its meaning. The Ministry of Education, Culture, Sports, Science and Technology (MEXT); Ministry of Health, Labour and Welfare; and Ministry of Agriculture, Forestry and Fisheries (MAFF) all promoted food and nutrition education from their respective standpoints. These separate efforts led to the establishment of the Liaison Council for the Promotion of Food and Nutrition Education in 2002. At that time, various issues relating to eating, food, and health were important concerns in Japan, including nutritional imbalance, individual and irregular diets, obesity and lifestyle-related diseases, food safety and food dependence on foreign countries, and the loss of Japan's traditional food (*washoku*) culture (Ministry of Agriculture, Forestry and Fisheries [MAFF], 2005, pp. 3–4). Consequently, the three ministries attempted to integrate their respective policies and legislations into food and nutrition education. In 2005, the *Shokuiku kihonhō* (Food and Nutrition Education Basic Act) was enacted. Simultaneously, MEXT created the "nutrition teacher" system to encourage the teaching of food education. However, the president of the Japan Association of Home Economics Education, Katsuko Makino (2003), argued that food education had already been incorporated into home economics education in elementary through upper secondary school curricula via the Course of Study (Japan's national curriculum standard). Makino (2003) also proposed that food content, such as cooking, consumer education, and life and culture, had been comprehensively taught in home economics. In addition, Makino (2003) argued that for food education to be effective, it should be taught by home economics teachers in collaboration with nutritionists (as opposed to nutrition teachers), physical education teachers, and classroom teachers. Consequently, she expressed skepticism about the creation of a nutrition teacher system (p. 13).

The Food and Nutrition Education Basic Act currently defines *shokuiku* as follows:

> It is required that *shokuiku* … be certainly positioned as the basis of a human life which is fundamental to intellectual education, moral education, and physical education, and be promoted for the purpose of educating people to become a person who is able to acquire knowledge about "food and nutrition" and [the] ability to choose appropriate "food and nutrition" for the person's own sake through their various experiences, which enables them to adopt healthy dietary habits.
>
> (MAFF, 2005, p. 2)

According to Reiher (2012), this definition of *shokuiku* is vague. The definition in the Food and Nutrition Basic Act is similar to that of Ishizuka in the sense that food and nutrition education is described as fundamental to intellectual and physical education; however, in the Basic Act, Ishizuka's "talent education" was changed to "moral education." The Food and Nutrition Basic Act stipulates the competencies that should be cultivated by food and nutrition education and covers different areas of education, such as both home and school education. Ministry of Education, Culture, Sports, Science and Technology (MEXT) (2019) maintains that while food education should be prioritized in home education, it should also be promoted by nutrition teachers in schools. In 2020, *shokuiku* was promoted as part of Japan's Sustainable Development Goals Action Plan; thereafter, the Fourth Basic Plan for the Promotion of *Shokuiku* was initiated in 2021.

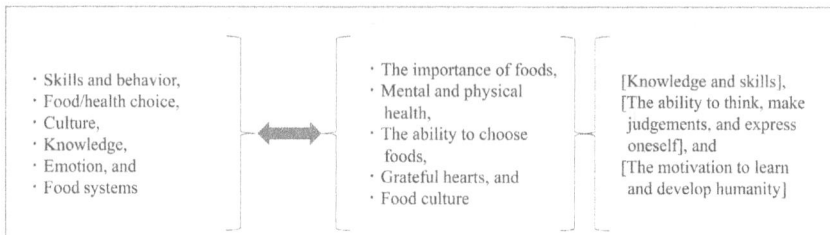

Figure 7.1 Components/themes of food literacy by Truman et al. (2017) and MEXT (2019).

What is the relationship between Shokuiku and food literacy?

According to Truman et al. (2017) and Omori and Kaneko (2021), there are several definitions of food literacy. Truman et al.'s (2017) interpretation is useful for the present research. They argue that current definitions of food literacy include six themes: (1) skills and behaviors that describe physical actions or abilities involving food; (2) food/health choices that involve actions associated with informed choices around food use; (3) culture that encompasses the societal aspects of food; (4) knowledge that refers to the ability to understand and seek information about food; (5) emotions that are influenced by attitudes and motivation; and (6) an understanding of the complexity of food systems, such as their environmental impact, food wastage, and food risk/safety.

MEXT produced *Shoku ni kansuru shidō no tebiki* (*The Handbook on Food Instructions*) as a guide in 2007, which was revised in 2010 and 2019 and describes six themes for food education: the importance of food, mental and physical health, the ability to choose food, having a grateful heart, social skills, and food culture. Each theme has three competencies, as described in the Course of Study for each subject: knowledge and skills; the ability to think, make judgments, and express oneself; and the motivation to learn and develop humanity. As Figure 7.1 shows, there are similarities between the definitions of food literacy provided by Truman et al. (2017) and MEXT: Both cover cognitive skills and proceed from the individual to the social level.

After analyzing the food literacy literature, Omori and Kaneko (2021) concluded that all children in Japan should develop food literacy through effective interactions between home economics education, school lunches, and *shokuiku*.

A brief history of food and nutrition education in home economics and school lunches

From an international perspective, McCloat and Caraher (2020) proposed that home economics "is ideally placed to utilise its pedagogical approaches and philosophical underpinning" (p. 320) for teaching food education. Similar to Makino (2003), Suzuki (2014) claimed that home economics teaches more about diet than other subjects in Japanese schools and recommended collaboration with other subjects and activities, with a focus on cooking practice.

Although home economics is important for teaching food and nutrition, Stitt (1996) argues that a comprehensive approach may be more effective. Following a literature review and online survey on food literacy, Pendergast and Dewhurst (2012) concluded that home economics teachers "should lead and co-ordinate in the development of and vision for food literacy" (p. 258). This led us to investigate the history of home economics (formerly *kaji*).

Kaji before World War II and home economics after WWII

Elementary schools were first established in Japan in 1872. From 1872 to 1945, compulsory education in the form of elementary schooling was reorganized several times. During that period, while sewing was always compulsory for girls, home economics, which taught food and nutrition in the modern sense, did not exist at the ordinary elementary level. From 1911 to 1918, higher elementary school was not compulsory; however, *kaji*, which can be translated as "home economics, homemaking, household, or domestic science," was part of the science classes taught to girls at that level. It included learning content that was more closely tied to science, such as food composition. In 1919, *kaji* was an independent, optional subject; however, in 1926, it became compulsory for girls in higher elementary schools. Knowledge and skills in food and nutrition were taught in this practical subject (MoE, 1980).

The Girls' High School Regulations were issued in 1895, although attendance was not compulsory. *Kaji* was taught as a compulsory subject from its establishment until the beginning of World War II. Girls' high schools taught advanced knowledge and skills in food and nutrition, rather than providing elementary schooling, and emphasized practice as well as theory. Additionally, chemistry, botany, and physiological hygiene, which were components of the science subject, included teaching on food and nutrition. Nevertheless, the relationship between *kaji* and science in practice remained unclear. In contrast, middle schools for boys did not historically teach *kaji* as a subject (MoE, 1980; Tokyo-Kaiseikan, 1937).

Ishizawa (1921, 1931), a professor of *kaji*, stated that it should be connected to other subjects such as science, mathematics, moral education, and civics and argued that science was necessary to rationalize empirical *kaji* and make it scientific. Ishizawa (1921, 1931) claimed that acquiring applied scientific knowledge and developing scientific attitudes and deductive reasoning was important in *kaji*. He intended to cultivate students' ability to improve their lives by learning *kaji*. Ishizawa and Tsunemi (1939) described the relationships between science subjects and the teaching areas/materials of *kaji* as follows. The subjects of botany, zoology, and mineralogy were related to raw clothing materials, food, housing materials, materials for tools, and fuel types. The physiology and hygiene subject covered physiology, nutrition, hygiene and disease, childcare, and nursing. Physics and chemistry taught knowledge related to the choice of clothing materials, dyeing, washing, food cooking, ventilation of dwellings, heating, and lighting (p. 69).

However, Kondo (1935), a professor of science education, acknowledged that science and *kaji* were closely related but expressed skepticism about Ishizawa's opinions. Kondo (1935) argued that science and *kaji* had their own specific positions and missions; therefore, there was no need to teach them together. As such, when creating chemistry textbooks for girls' high schools, Kondo and Takeshima (1936) omitted the content on vitamins that was commonly included in these textbooks to avoid unnecessary duplication with *kaji*. However, it is premature to assume that Kondo shunned *kaji*. As mentioned above, he was well aware that science and *kaji* each had their own specific purpose. Kondo published numerous textbooks for girls' high schools on *kaji* (e.g., Kondo, 1930) and science. It can be, thus, inferred that Kondo simply intended to prevent duplication between the subjects.

Why were food and nutrition taught only to girls before WWII? The answer to this question can be clearly found in the Minister's statement in 1899, which declared that girls' high schools should aim to develop "good wives and wise mothers" (MoE, 1980, p. 119). Such government policy was used to cultivate loyal nationalists and were not easily influenced by rational and individualistic Western cultures (Shibukawa, 1971).

After WWII, Japan's schooling system was reorganized under the influence of the United States, and democratic education was introduced. The schooling system became coeducational and consisted of six compulsory years of elementary school and three years of lower secondary schooling. In grades five and six, a newly formed home economics subject, which partially included food and nutrition education, was made compulsory for both girls and boys. In lower secondary schools, home economics was also newly introduced as vocational education in 1947, based on discussions with and recommendations from the Civil Information and Education Section of the General Headquarters (e.g., Hônoki, 1988a, 1988b). However, during Japan's rapid economic growth, the Course of Study for Lower Secondary Schools was revised in 1958. This revision influenced home economics in two ways. First, home economics was integrated with technology and established as a new compulsory subject called "technology and home economics." Second, home economics also became an independent elective subject. Both of these newly established subjects included knowledge and skills related to food and nutrition. Furthermore, two types of content were taught: technology for boys and home economics for girls (later, boys studied some home economics, while girls also studied some technology). After the revision of the Course of Study for Lower Secondary Schools in 1977, which aimed to achieve the harmonious intellectual, moral, and physical development of students, home economics as a single subject was no longer offered as an elective subject (MoE, 1977). Drawing on Slater's three definitions of food literacy (functional, interactive, and critical literacy (Slater, 2013, p. 623)), Kawamura (2019) pointed out that postwar food education within home economics, especially during the period when only girls pursuing their in secondary education were enrolled for this subject, has traditionally been characterized by both functional and interactional literacy.

According to the 1989 revision of the Course of Study for Lower Secondary Schools, the same contents of technology and home economics were to be learned by both boys and girls. In addition, home economics, previously learned only by girls, also became a compulsory subject for *both* boys and girls through the Course of Study for Upper Secondary School revised in the same year. Home economics is now firmly rooted in the school curriculum from elementary to upper secondary schools as a subject studied by both girls and boys and taught by both female and male teachers. This situation reflects one of the Japanese government's policies to realize a gender-equal society. Since 1947, home economics has covered knowledge and skills in food and nutrition; however, its position in the school curriculum has shifted due to social and cultural changes, such as increasing gender equality and the nuclearization of the family. In the newly revised Course of Study for Lower Secondary Schools released in 2017, the technology and home economics subject includes learning about clothing, food, and housing; the role of diet and nutritional characteristics of students; the necessity of meals that meet students' nutritional needs; and daily food preparation and local food culture (MEXT, 2017). Consequently, home economics education in Japan has developed critical food literacy, as well as both functional and interactive food literacy, from the periods when both genders were required to study home economics in lower secondary schools and, later, in upper secondary schools.

While home economics—*kaji* before WWII—historically played an important role in *shokuiku*, it also had a gender bias. Further, it was not always a compulsory subject like Japanese and mathematics but was sometimes an elective subject or positioned as vocational preparatory education. Thus, the teaching of home economics has been influenced by curriculum politics and social changes throughout history.

An introduction to the nutrition teacher system

In 1954, the School Lunch Law was enacted. According to Kawagoshi and Suzuki (2014), the introduction of school lunches originated from social and charitable projects to assist disadvantaged children. Subsequently, as the tradition of school lunches began to spread, the aim shifted from encouraging students to attend school to improving nutrition. The process of enacting laws, establishing financial resources, and providing school lunches also became part of national state policy. After Japan's defeat in WWII, school lunches were initiated with the support of the Allies to improve students' physical condition. Nutritional education was also provided, although no certification for teaching nutrition was introduced. Certification for nutrition teachers began in 2005, and they currently have two roles in schools: first, to teach nutrition, and second, to manage school lunches. Both roles should be promoted as integral parts of creating a *shokuiku* program in schools. However, compared with teachers of other subjects, nutrition teachers are not placed in all schools; whether they are placed is determined by the local education board.

Discussion: a strategy for *Shokuiku* in schools

As a part of science, engineering, technology, (liberal) arts, and mathematics (STEAM) education, *shokuiku* can provide a holistic understanding of how knowledge, skills, and perspectives can be integrated and applied to solve complicated and real-world problems. From a historical perspective, home economics teachers can play a vital role in teaching *shokuiku*. The enactment of the Food and Nutrition Education Basic Act is the first step toward enabling home economics teachers to take the initiative to engage in *shokuiku* in STEAM education.

Several ministries were involved in passing the Food and Nutrition Education Basic Act. MEXT, in particular, promoted *shokuiku* in schools. More specifically, after the enactment of the Food and Nutrition Education Basic Act, MEXT produced *The Handbook of Food Instructions* as guidance for schools. According to this handbook (MEXT, 2019), *shokuiku* should mainly be taught by nutrition teachers; however, the handbook states that it should be considered in other contexts as well, such as subject teaching, lunchtime, and student-led activities.

The handbook describes the objectives of food instruction, which comprise three competencies, as follows:

Knowledge and skills: Students understand the importance of food, nutritional balance, and food culture and acquire knowledge and skills related to healthy eating habits.

The ability to think, make judgments, and express oneself: Students develop the ability to manage and make decisions about their diet and food choices based on accurate knowledge and information.

The motivation to learn, and develop humanity: Students proactively recognize the importance of a healthy diet for themselves and others; develop an appreciation for people involved in food, food culture, food production, and other areas; and foster the ability to build relationships by understanding table manners and engaging in opportunities to eat with others.
(MEXT, 2019, p. 16, translated by the authors)

The second revised edition of the handbook reorganized six themes and perspectives on *shokuiku* as follows:

The importance of food: Students understand the importance of food and the pleasure and enjoyment of eating.

Mental and physical health: Students understand the nutrition and dietary practices that are desirable for the maintenance and promotion of mental and physical growth and health and acquire the ability to manage their diets.

Ability to choose foods: Students acquire the ability to make their own judgments about food quality and safety based on accurate knowledge and information.

Grateful hearts: Students value food and express their gratitude to those involved in food production.

Social skills: Students gain the ability to build relationships by understanding eating habits and opportunities.

Food culture: Students understand and respect each region's products, food culture, and food-related history.

(MEXT, 2019, p. 16, translated by the authors)

Each theme consists of three competencies, similar to the Course of Study for each subject. Elementary and lower secondary school teachers organize *shokuiku* programs using the handbook and other information that is provided by various ministries such as MAFF.

Drawing on the ideas of Wiggins and McTighe (1998), Drake and Burns (2004) proposed the KNOW/DO/BE bridge to design and assess integrated curricula. The "KNOW" aspect asks, "What is most important for students to know?," the "DO" asks, "What is it most important for students to be able to do?," and the "BE" asks, "What kind of people do we want students to be?" When designing and assessing the *shokuiku* program in Japan, the KNOW/DO/BE bridge model can be fully applied to the three above-mentioned competencies. According to Drake and Burns (2004), "KNOW" and "DO," the substructure of the bridge, interact to support "BE" as the superstructure of the bridge; thus, Japan's three competencies can be regarded as a bridge model.

Since the Food and Nutrition Education Basic Act was enacted, teaching *shokuiku* in the form of a single subject such as home economics has been found difficult. Although teachers, especially secondary school teachers, may be aware of the barriers and challenges to cross-curricular approaches, it is also necessary to consider teaching *shokuiku* in a wider range of subjects in collaboration with other streams. The issuance of the certification for nutrition teachers has begun only recently. As MEXT (2019) argues, teachers of related subjects must cooperate to promote *shokuiku* as a whole. The importance of STEAM education has been highlighted in Japan (Ministry of Education, Culture, Sports, Science and Technology [MEXT] (Central Council for Education), 2021), and *shokuiku* could be taught as part of STEAM education. For example, science teaching can include knowledge and skills related to nutrition, whereas mathematical skills, such as measuring, can be used in food education. The Food and Nutrition Education Basic Act draws attention to Japan's traditional food culture, which can be taught in social studies classes. These aspects overlap with those of home economics. Therefore, it is crucial that nutrition teachers and teachers of subjects related to STEAM education cooperate closely to achieve the goals of *shokuiku* in schools.

Drawing on Drake and Burns' models of an integrated curriculum (2004), Vasquez et al. (2013) proposed a continuum of science, technology, engineering, and mathematics (STEM) approaches: disciplinary, multidisciplinary, interdisciplinary, and transdisciplinary. Drake and Burns (2004) pointed out that "the Japanese are turning to [an] integrated curriculum" (p. 4). In this context, *shokuiku* can be taught through integrated curricula, using both multidisciplinary and interdisciplinary approaches. Drake and Burns (2004) and Vasquez et al. (2013) suggested a transdisciplinary approach to STEM education and the integrated curriculum. In Japan, integrated learning activities are known as the "period for integrated studies" in elementary and lower secondary schools and the "period for inquiry-based cross-disciplinary study" in upper secondary schools. A transdisciplinary approach can be adopted through these activities. However, while the objectives, teaching contents, and teaching methods of each subject are strictly outlined in the Courses of Study, these integrated learning activities are established by each school. This emphasizes the professional judgment of teachers in schools and fosters originality and ingenuity. Consequently, the entire school must work together to teach *shokuiku* comprehensively across subjects, rather than as a single subject. Hence, teachers must build effective collaborative relationships, primarily with home economics and nutrition teachers, under the leadership of senior management groups in schools.

Considering the history of *shokuiku* and educational trends in both Japan and worldwide, we propose two approaches to *shokuiku*: multidisciplinary and interdisciplinary. These sample models are primarily targeted at lower secondary schools but can also be applied to elementary schools that have introduced a subject-teacher system. Of course, even in elementary schools that have introduced a classroom-teacher system, teachers can connect the three competencies in each subject area by carefully interpreting the Course of Study/national curriculum standards and the MEXT handbooks and consulting with colleagues. Figures 7.2 and 7.3 illustrate the proposed

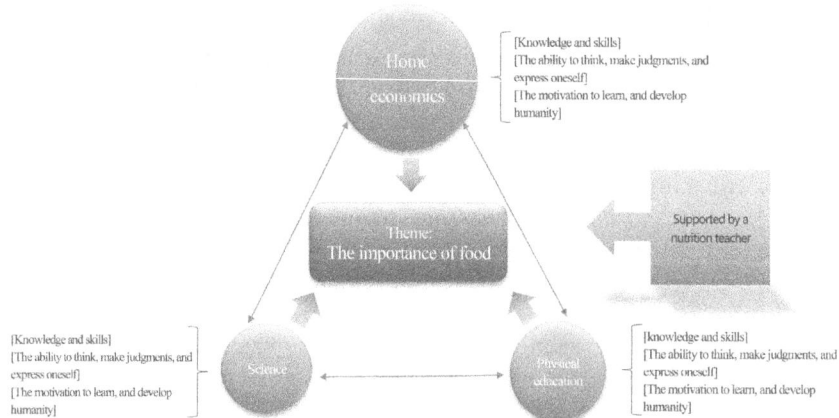

Figure 7.2 A model sample of the multidisciplinary approach.

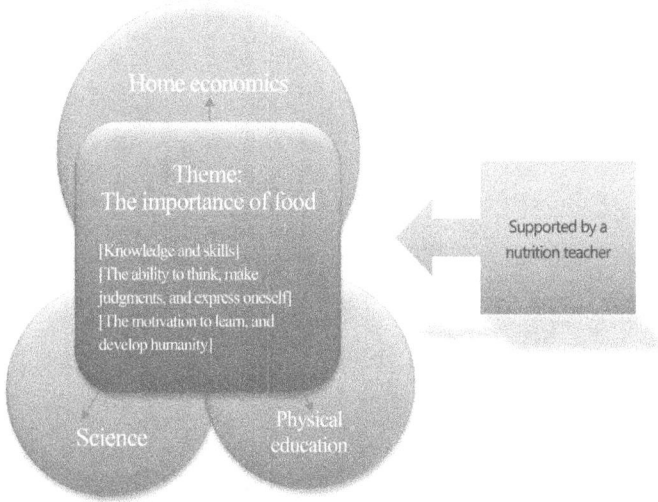

Figure 7.3 A model sample of the interdisciplinary approach.

multidisciplinary and interdisciplinary approaches, respectively. Table 7.1 summarized the details of the two approaches (see page 110). Although each subject is independent and accesses the themes of *shokuiku* through a multidisciplinary approach, collaboration among related subjects is essential for the effective promotion of *shokuiku*.

Conclusion

Shokuiku is both an old and new educational issue and policy affected by changing social and cultural conditions. This study aimed to examine the development of food literacy. Since the beginning of this millennium, health issues such as lifestyle-related diseases and global concerns related to food self-sufficiency have led Japan's government to emphasize *shokuiku*. This has also been a strategic concern for the nation in relation to human development in the STEM fields. Historically, home economics—known as *kaji* before WWII—has undoubtedly played a crucial role in *shokuiku*. The home economics subject considers the context of daily life, allowing all students from elementary through upper secondary schools to learn how to improve their lives. However, as *shokuiku* covers a wide range of topics, it is not easy to teach them all within the limited number of instruction hours assigned to home economics. To address this issue, the nutrition teacher education system has also been institutionalized.

Japan has employed a separate subject curriculum in schools since the mid-19th century. Although integrated learning, such as the period for integrated studies, has had a firm place in the school curriculum since the beginning of the millennium, its relationship with individual subject studies remains unclear and complex.

Table 7.1 Two approaches for *Shokuiku* instruction.

		Multidisciplinary approach	*Interdisciplinary approach*
Organizing center		Course of Study (National Curriculum Standards) which includes three competencies: [Knowledge and skills], [The ability to think, make judgments, and express oneself], and [The motivation to learn, and develop humanity] of each subject organizing around theme.	*The Handbook on Food Instruction* by the MEXT (the second revised edition in 2019), and three competencies of related each subject, such as home economics, science, and physical education.
Development of three competencies		Three competencies are learned through the structure of each subject.	Under the common theme, *shokuiku-* interdisciplinary three competencies are firstly stressed, and subjects are connected.
Instruction		Inquiry or PBL, Experimental learning, Design thinking, etc.	These approaches require students to carry out investigations, analyze and interpret data, and communicate and work with others.
Role of teachers	Before lessons	Teachers collaboratively organize the curriculum around common learning across subjects based on collaboration with nutrition teachers. *This collaboration is very important for teachers at secondary schools. Generally, they may become aware of barriers or challenges to cross curricular approaches, such as multidisciplinary and interdisciplinary approaches.*	
	In lessons	Teachers lead instruction and carefully facilitate student learning in each subject or across subjects.	
	After lessons	Teachers collaboratively reflect the curriculum and give a feedback to students as well as other teachers. *Like lesson study, "reflection" with colleagues in schools is a very important opportunity to share ideas and students' understanding, and take the next strategy.*	

Note: Drake and Burns (2004) and Vasquez et al. (2013) proposed three integrating approaches: multidisciplinary, interdisciplinary, and transdisciplinary. However, in Japan, there are integrated learning activities known as the "period for integrated studies" in elementary and lower secondary schools, and the "period for inquiry-based cross-disciplinary study" at upper secondary schools. Transdisciplinary approach is used in these learning activities; therefore, we omitted it in this table.

Source: Drake, S. M., & Burns, R. C. (2004). *Meeting standards through integrated curriculum.* Association for Supervision and Curriculum Development. Vasquez, J. A., Snider, C., & Comer, M. (2013). *STEM lesson essentials, grades 3–8.* Heinemann. Ministry of Education, Culture, Sports, Science and Technology (MEXT) (2019). *Shoku ni kansuru shidō no tebiki: Dainiji kaitei-ban (The handbook on food instruction: 2nd revised edition).* MEXT [in Japanese].

Shokuiku enables students to learn about real-world problems. Considering Japan's specific educational traditions and trends, we argue that *shokuiku* should be introduced as a part of STEAM education (the "A" here can refer to a wide range of subjects from fine to liberal arts) within the common curriculum and for every student. We also propose that home economics and nutrition teachers should engage with teachers of other subjects, primarily science, physical education, and social studies teachers, to implement multidisciplinary or interdisciplinary approaches. As such, successful *shokuiku* depends on building collaborative relationships among schoolteachers.

From a historical perspective, Jackson et al. (2020) highlighted that to provide authentic learning, more empirical research is necessary to investigate which approaches—including interdisciplinary, multidisciplinary, and transdisciplinary—are suitable for students at the primary and secondary education levels in the integrated curricula. This study can support this point. Additionally, we argue that both pre- and in-service teacher education needs to provide more opportunities to learn about the integrated curriculum of STEAM education as well as to engage in professional studies, and specialize in subject education.

Acknowledgment

This work was supported by the Japan Society for the Promotion of Science KAKENHI Grant Numbers JP22K02982, JP21H00919 (JP23K20744).

Note

Part of this study was presented at the Annual Meeting of the American Educational Research Association in Chicago in April 2023.

References

Drake, S., & Burns, R. (2004). *Meeting standards through integrated curriculum.* Association for Supervision and Curriculum Development.

Hônoki, K. (1988a). Amerika-gawa-shiryō yori mita kateika no seiritukatei (1): Kaji-ka, saihō-ka no tougou no kettei (The formation of home economics education in reference to CIE documents (1) – From December 1945 to August 1946). *Journal of the Japan Association of Home Economics Education, 30*(3), 35–40 [in Japanese with English abstract].

Hônoki, K. (1988b). Amerika-gawa-shiryō yori mita kateika no seiritukatei (2): Kasei-kyōikukatei-kaisei-iinkai no seiritsu (The formation of home economics education in reference to CIE documents (2) – From December 1945 to August 1946). *Journal of the Japan Association of Home Economics Education, 30*(3), 41–47 [in Japanese with English abstract].

Ishizawa, Y. (1921). *Kaji-kyōjyuhō (The teaching method of kaji)* (3rd ed.). Shuseido [in Japanese].

Ishizawa, Y. (1931). *Saishin kaji-kyoujyuhō seigi (New teaching method of kaji)*. Kyōikukenkyūkai [in Japanese].

Ishizawa, Y., & Tsunemi, I. (1939). *Saishin kaji-kyōjyuhō (New teaching method of kaji-ka)*. Soubunsha [in Japanese].

Ishizuka, S. (1896). *Kagakuteki shokuyō chōjyu ron (A chemical/scientific theory of nutrition for health and longevity)*. Hakubunkan [in Japanese]. Retrieved from https:// dl.ndl.go.jp/pid/836793/1/

Jackson, C., Tank, K. M., Appelgate, M. H., Jurgenson, K., Delaney, A., & Erden, C. (2020). History of integrated STEM curriculum. In C. C. Johnson, M. J. Mohr-Schroeder, T. J. Moore, & L. D. English (Eds.), *Handbook of research on STEM education* (pp. 169–183). Routledge.

Kawagoshi, Y., & Suzuki, K. (2014). Gakkō-kyūshoku no yakuwari to kouka (1) Sengo no gakkō-kyūshoku-hō seitei made no keika ni tuite (Role and effect of the Japanese school lunch program 1: Progress up until the establishment of the postwar school lunch act). *Bulletin of Seinan Jo Gakuin University, 18*, 129–138 [in Japanese with English abstract].

Kawamura, M. (2019). Shougai ni wataru kiso to naru shokuiku-seikatsu: Fūdo riterashi no kanten kara (Food education as a long-life foundation: From the perspective of the development of food literacy). *Journal of the Japan Association of Home Economics Education, 62*(2), 107–111 [in Japanese].

Kondo, K. (1930). *Shinpen kaji-kyōkasho (New Kaji-ka textbook)* (revised 4th ed.). Kōfūkanshoten [in Japanese].

Kondo, K. (1935). Kaji to rika no kankei (The relationship between *kaji* and science). In *Shihan-kyōiku rika kyōiku kōza (The Normal University Science Education Courses)* (Vol. 13, pp. 1–7). Kenbunsha [in Japanese].

Kondo, K., & Takeshima, S. (1936). *Jyoshi rika kagaku kyōkasho (Girls' high school science: Chemistry textbook)*. Meguroshoten and Seibido (revised 8th ed.) [in Japanese].

Lawton, D. (1980). *The politics of the school curriculum*. Routledge & Kegan Paul.

Makino, K. (2003). Ronten: Shokuiku wa kateika jyūjitu kara (Issue: Food education starts with improved home economics). *Yomiuri Shimbun* (newspaper), 24th October, p. 13.

McCloat, A., & Caraher, M. (2016). Home economics as a food education intervention: Lessons from the Irish secondary education context. *Education and Health, 34*(4), 104–110.

McCloat, A., & Caraher, M. (2020). An international review of second-level food education curriculum policy. *Cambridge Journal of Education, 50*(3), 303–324.

Ministry of Agriculture, Forestry and Fisheries (MAFF) (2005). Basic act on *shokuiku* (food and nutrition education) (Act No. 63 of June 17, 2005). Retrieved from: http://www.maff.go.jp/e/policies/tech_res/attach/pdf/shokuiku-19.pdf

Ministry of Education, Culture, Sports, Science and Technology (MEXT). (2017). *Chūgakkō gakushū-shidō yōryō (The course of study for lower secondary school)*. MEXT [in Japanese]. Retrieved from https://erid.nier.go.jp/files/COFS/h29j/index.htm

Ministry of Education, Culture, Sports, Science and Technology (MEXT). (2019). *Shoku ni kansuru shidō no tebiki: Dainiji kaitei-han (The handbook on food instructions: 2nd revised edition)*. MEXT [in Japanese]. Retrieved from https://www.mext. go.jp/a_menu/sports/syokuiku/1292952.htm

Ministry of Education, Culture, Sports, Science and Technology (MEXT) (Central Council for Education). (2021). *"Reiwa no nihongata-gakkōkyōiku" no kouchiku wo mezashite: Subete no kodomotachi no kanousei wo hikidasu, kobetusaitekinamnabi to kyōdoutekina manabi no jitugen (tōshin) (Toward the construction of "Japanese style* school education": Optimal individualized learning and collaborative learning that bring out the potential all children (report)). MEXT [in Japanese]. Retrieved from https://www.mext.go.jp/content/20210126-mxt_syoto02-000012321_2-4.pdf

Ministry of Education, Science and Culture (MoE) (1977). *Chūgakkō gakushū-shidō yōryō (The course of study for lower secondary school)*. Ministry of Finance [in Japanese].

Ministry of Education, Science and Culture (MoE) (1980). *Japan's modern educational system: A history of the first hundred years*. Ministry of Finance.

Morita, M. (2004). Shokuiku no haikei to keii: "Shokuiku kihonhōan" ni kanrenshite (Background and history of food and nutrition education: Focusing on the bill "Food and Nutrition Education Basic Act"). *Issue Brief*, 457, 1–10 [in Japanese].

Omori, K., & Kaneko, K. (2021). Fūdo riterashi ni kansuru kenkyū no kokusaiteki-dōkō (International trends in food literacy studies). *Journal of the Japan Association of Home Economics Education*, 72(4), 206–217 [in Japanese with English abstract].

Pendergast, D. (2012). The intention of home economics education. In D. Pendergast, S. McGregor, & K. Turkki (Eds.), *Creating home economics futures: The next 100 years* (pp. 12–23). Australian Academic Press.

Pendergast, D., & Dewhurst, Y. (2012). Home economics and food literacy: An international investigation. *International Journal of Home Economics*, 5, 245–263.

Reiher, J. (2012). Food pedagogies in Japan: From the implementation of the basic law on food education to Fukushima. *Australian Journal of Adult Learning*, 52(3), 507–531.

Shibukawa, H. (1971). An education for making good wives and wise mothers. *Education in Japan*, 6, 47–57.

Slater, J. (2013). Is cooking dead? The state of home economics food and nutrition education in a Canada province. *International Journal of Consumer Studies*, 37(6), 617–624.

Stitt, S. (1996). An international perspective on food and cooking skills in education. *British Food Journal*, 98(10), 27–34.

Suzuki, Y. (2014). Food and nutrition education in home economics education. *Journal of the Japan Association of Home Economics Education*, 57(3), 238–241.

Tokyo-Kaiseikan (1937). *Kōtōjyogakkō kyōiku hōrei yōran (Directory of education laws and regulations for the girls' high school)*. Tokyo-Kaiseikan (3rd revised ed.) [in Japanese].

Truman, E., Lane, D., & Elliott, C. (2017). Defining food literacy: A scoping review. *Appetite*, 116, 365–371. https://doi.org/10.1016/j.appet.2017.05.007

Vasquez, J., Sneider, C., & Comer, M. (2013). *STEM lesson essentials, grades 3–8: Integrating science, technology, engineering, and mathematics*. Heinemann.

Wiggins, G., & McTighe, J. (1998). *Understanding by design*. Association for Supervision and Curriculum Development.

World Health Organization (WHO). (2004). *Global strategy on diet, physical activity and health*. WHO. Retrieved from https://iris.who.int/bitstream/handle/10665/43035/9241592222_eng.pdf?sequence=1

8 Pre-service teacher education for STEAM activities that combine data modeling and design processes

Takashi Kawakami and Keiichi Nishimura

Introduction

In recent years, research has been conducted worldwide on STEAM (science, technology, engineering, the liberal arts, and mathematics) education, which refers to finding and solving real-world problems using concepts and skills from mathematics, science, and the liberal arts, while incorporating design methodologies of engineering and using the appropriate technologies (Moore et al., 2020). In Japan, STEAM education (including STEM [science, technology, engineering, and mathematics] education) is emphasized as a form of cross-curricular learning (Ministry of Education, Culture, Sports, Science and Technology [MEXT] (Central Council for Education), 2021). To implement STEAM education in Japanese school curricula, it is important to explore STEAM education and pre-service teacher (PST) education based on domestic educational and social contexts (Isozaki & Isozaki, 2021). Additionally, teachers must understand the nature and learning values of each STEAM subject and domain (Isozaki & Isozaki, 2021). For example, G1–G12 mathematics (which combine mathematics and statistics) play a pivotal role in STEAM education. However, the incorporation of mathematics into STEAM classes is limited to performing calculations and using symbols; therefore, its role in STEAM education is weak (e.g., English, 2017; Maass et al., 2019). Moreover, it is challenging for teachers in Japan and other countries to develop and implement STEAM lessons that foreground the role of mathematics (e.g., Kawakami & Saeki, 2021; Nishimura & Tachikawa, 2019; Tytler et al., 2019). Accordingly, strategies for PSTs to design and implement STEAM lessons that consider the nature and learning value of each STEAM subject and domain need to be further explored.

In this chapter, we aim to address this research gap and examine PST education for STEAM education from the perspective of mathematics education by incorporating *data modeling* (Kawakami, 2022; Lehrer & English, 2018) and *design processes* (e.g., English, 2017) that highlight the role of each STEAM subject and domain, with a focus on mathematics and

DOI: 10.4324/9781003392545-9

engineering. At the primary and secondary levels, STEM education practices that combine data modeling and design processes are being developed (e.g., English, 2017). However, to the best of our knowledge, no previous study has explored a STEAM education approach that combines data modeling and design processes at the PST education level. To this end, we examine the Japanese context to present an example of PST education for STEAM activities that combine complementary data modeling and design processes, from which implications are drawn regarding STEAM PST education.

Challenges of STEAM education in Japan

STEAM education

In Japan, STEAM education is defined as a cross-curricular educational approach that enables students to apply what they learn in each STEAM subject and domain to find and solve real-world problems (MEXT (Central Council for Education), 2021). The scope of liberal arts incorporates not only arts and culture, but also living environment, economics, law, politics and ethics, among others. In a systematic review of STEM education, Moore et al. (2020) identified the use of real-world problems and the linking of skills, practices and concepts across STEM subjects and domains as the characteristics of integrated STEM education. Variations in the integration of STEM education have also been reported (Moore et al., 2020). For example, Drake and Burns (2004) proposed three approaches to the integration of subject areas: (1) concepts and competencies are learned separately in each subject and domain, grounded on a common theme (*multidisciplinary*); (2) concepts and skills from two or more subjects and domains are linked (*interdisciplinary*); and (3) concepts and skills from two or more subjects and domains are applied to solve real-world problems or projects (*transdisciplinary*).

Regarding the relationship between STEM and mathematics education, Anderson et al. (2020) categorized the approaches to STEM education adopted by mathematics education researchers into three types, based on the utilization of STEM contexts to (1) engage students in learning; (2) develop students' problem-solving and critical thinking skills; and (3) promote students' understanding of mathematics. Building on Anderson et al. (2020), Kawakami and Saeki (2021) identified three goals of STEM education with mathematics at its core: (1) learning interdisciplinary content in STEM disciplines; (2) developing generic competencies for interdisciplinary problem solving; and (3) understanding mathematical and statistical concepts. These goals are not independent, while multiple goals may be pursued in a single lesson or unit. This categorization of goals also applies to STEAM education.

Practicalities of STEAM teacher education in Japan

The Courses of Study in Japan, which are general standards for primary and secondary schooling, have been revised approximately every 10 years. The curriculum guidelines for primary and lower secondary schools were revised in 2017, and those for senior high schools were revised in 2018. In particular, "basic inquiry-based study of science and mathematics" and "inquiry-based study of science and mathematics" were established as new subjects for senior high schools to promote cross-curricular learning centered on science and mathematics. The Ministry of Economy, Trade and Industry has also been promoting the "Future Classrooms" demonstration project since 2018 to present a new learning method that uses one computer terminal per student, and various educational technology projects for these new subjects. One such project is the STEAM Library (https://www.steam-library.go.jp), which aims to encourage a cyclical learning process of "knowing" and "creating" (drawing upon students' enthusiasm), and promote the "STEAMization of learning." As of August 2022, 173 educational content modules have been implemented.

The following is an overview of the STEAM educational material "Wind Power" (https://www.steam-library.go.jp/content/75), which involves data, mathematics, and design activities mainly intended for junior high and high school students. The material comprises five 50-minute lessons, each with video content, lesson slides, worksheets, and lesson plans. For example, in Lesson 2, a wind turbine model is used to develop and test hypotheses on how variables, such as the number and length of turbine blades, blade angle, fan strength, and fan angle affect power generation. Each lesson poses a "key question," as follows:

Lesson 1: Using wind turbines
 How do we meet our own energy needs, and how do these needs affect the natural world?
Lesson 2: Wind turbine design
 What factors affect the output of a wind turbine?
Lesson 3: Problem with wind power
 What are the challenges of using conventional wind turbines in Japan, and how should you develop your argument to make your point?
Lesson 4: Transition to offshore power generation
 What are the different disadvantages of onshore vs. offshore wind turbines? What do you think should be done to overcome the disadvantages of offshore wind turbines?
Lesson 5: Wind power breakthrough
 Could wind power be a breakthrough for Japan?

The content presented above is aimed at a wide range of grade levels. As lesson plans are limited to show the flow of the lesson, teachers need to determine what they do in the classroom. This part can be challenging, as middle and high school mathematics and science teachers in Japan have little experience in designing cross-curricular lessons that emphasize how each STEAM subject and domain is relevant in mathematics and science lessons. To aid teachers, we created a "Room for Future Classroom" on a website where teachers can post lesson plans they have implemented using content from the STEAM Library website, comment on the content, and receive feedback from teachers in the field. However, teachers' use of STEAM Library content in the classroom has been limited. Therefore, mathematics teachers must engage in interdisciplinary activities themselves, which could enable them to design STEAM lessons that highlight the nature and learning values of each STEAM subject and domain, with a focus on mathematics, as part of their PST education and professional development.

STEAM activities combining data modeling and design processes

We investigated the data modeling and design processes that could be utilized to achieve successful STEAM education, which highlight the role of each STEAM subject and domain. This section describes the content of the data modeling and design processes, and discusses why they are necessary for STEAM education.

Data modeling process

Modeling comprises both real-world problem-solving and knowledge-development processes. As generating, validating, and revising models are universal across STEM disciplines (Hjalmarson et al., 2020), these processes are also essential to the STEAM education structure (English, 2017; Maass et al., 2019). Furthermore, data are numbers with a real-world context (Cobb & Moore, 1997), and involve concepts and content (variability) that incorporate mathematics and other STEAM subjects and domains, which provide cross-curricular and interdisciplinary contexts (Watson, Fitzallen, & Chick, 2020). This study focused on data modeling in school mathematics (e.g., Kawakami, 2017, 2022; Lehrer & English, 2018).

According to Hestenes (2010), a model is a representation of the structure (set of relationships between objects) of a given system (set of related objects). Data modeling generally involves posing an exploratory problem about a real-world situation, translating that problem into relevant data, and creating a model of variation to facilitate making predictions and estimations about the situation (Lehrer & English, 2018). Models are constructed, represented, and revised within a discipline so that they can be shared with others or adapted and applied to structurally similar situations (Lesh & Doerr, 2003). Specific models produced through data modeling are representations of the structure of data and stochastic variability, such as tables, graphs, and statistics (Lehrer & English, 2018). The data modeling process consists of the following five phases (Lehrer & English, 2018, p. 235).

DMP(a): Posing statistical questions within meaningful contexts that highlight variability.
DMP(b): Generating, selecting, and measuring attributes that vary according to the questions posed.
DMP(c): Collecting first-hand data so that learners can make decisions about the design of the investigations.
DMP(d): Representing, structuring, and interpreting samples and sampling variability.
DMP(e): Making informal inferences based on these processes.

The five phases do not necessarily proceed sequentially from the DMP(a) phase to the DMP(e) phase but are interrelated, and it is possible to move back and forth between phases. In the DMP(a) to (d) phases, the main goal is to understand the variability of the sample at hand, particularly in the DMP(d) phase, where the creation of models representing sample variability is actively undertaken. In the DMP(e) phase, the goal is to use these models to make predictions and estimations for unknown samples and populations. Decision-making and explanations to others are also emphasized in this phase.

More recently, the data modeling process has been incorporated into STEM education from primary school to teacher education. For example, Watson, Fitzallen, English, et al. (2020) analyzed through a data modeling process in a STEM-related context how third graders' (aged 8–9) conceptions of variability develop. In addition, English (2017) illustrated that linking data modeling and engineering design can promote STEM learning at the primary and secondary levels. At the teacher education level, Petrosino (2016) analyzed in-service teachers' use of data, measurement and data modeling through STEM teacher education. Similarly, Smith et al. (2019) implemented practices for PSTs to promote the understanding of STEM content through statistical problem-solving, including the data modeling process. Thus, the data modeling process is expected to promote real-world problem-solving and conceptual understanding in data-rich situations.

Design process

The design process has also been receiving increasing attention as the key to achieving STEAM education, due to its interdisciplinary nature (e.g., English et al., 2020; Yata et al., 2020). It typically involves problem scoping, idea generation, design and creation, testing and reflection, redesign and re-creation, and communication, to achieve optimal design (English et al., 2017, 2020). Traditionally, design activities have been emphasized in technology and engineering education (English et al., 2020). Today, design activities include the development of all forms of improvement-focused interventions, such as the provision of services, application of procedures and strategies, and development of policies (English, 2020). The design process in STEAM learning consists of seven phases (English et al., 2020, p. 78).

DP(a): Problem framing or scoping, during which problem boundaries, goals, and constraints are identified.

DP(b): Generation of multiple ideas for planning and potential construction, in contrast to focusing on developing a single idea.

DP(c): Balancing benefits and trade-offs when considering design ideas.

DP(d): Designing and constructing, including sketching and transforming a design into a product.

DP(e): Testing and reflecting on outcomes, including checking problem-goal(s) attainment and adherence to constraints.

DP(f): Redesigning and reconstructing; reflections on the initial design help identify improvements, with a subsequent design being developed.

DP(g): Reflecting on and communicating the overall design processes applied.

Similar to the data modeling process, the design process does not necessarily proceed in one direction; its components are interrelated, with the possibility of moving back and forth between phases. Thus, the goal of a sequence of activities is to create an optimal design. In the DP(d) and DP(e) phases, models are created and verified; these are mainly prototypes/scale models. Representations of the models' designs (called "design sketch models" in this chapter) are also produced, evaluated, and revised. The design process requires the use of cross-disciplinary knowledge, such as mathematics and science, as well as representations, content, and methods that involve multiple domains. This process also facilitates the development of such interdisciplinary knowledge (English et al., 2020). Another characteristic of the design process is the consideration of trade-offs in the pursuit of an optimal design (English, 2020); it involves both utilizing and constructing interdisciplinary knowledge.

The design process has also been incorporated into STEM education starting from primary school to teacher education (English et al., 2020). For example, English (2020) described STEM integration that adapted a design process for Grade 5 (10 years old) students. Fan and Yu (2017) demonstrated the application of integrative STEM understanding and higher-order thinking skills in high school students (16 to 17 years old) through a STEM approach that embeds engineering design practices. Regarding teachers, Koehler et al. (2007) demonstrated in-service teachers' integration of knowledge of content, pedagogy, and technology with an authentic design task involving the design and redesign of a fifth-grade website on the solar system. Thus, the design process is expected to promote the use and organization of interdisciplinary knowledge, and the development of generic competencies across STEM disciplines.

Why combine data modeling and design processes for STEAM education?

Table 8.1 presents the correspondence between the data modeling and design process phases, with direct equivalents in the same table row. The two processes can be seen as having complementary rather than contradictory emphases. On the one hand, the data modeling process underscores informal inference (i.e., prediction and estimation) and the generation, evaluation, and revision of models of data variability. On the other hand, the design process emphasizes design optimization, trade-offs and the generation, evaluation, and revision of prototypes/scale models and design sketch models. Both processes aim at future-oriented problem-solving to achieve an ideal or desired reality, such as prediction, estimation, and design optimization, considering the needs, uses, and limitations in the complex real world.

Furthermore, in both the data modeling and design processes, STEAM knowledge can be used or developed in an integrated manner through model generation, evaluation, and revision. Both processes can clarify the role of each STEAM subject and domain. For example, in the data modeling process, the roles of mathematics (e.g., statistical and mathematical knowledge) and technology (e.g., knowledge of the use of software for data analysis) within STEAM disciplines are particularly evident (Lehrer & English, 2018), and the data also provide an interdisciplinary context (Watson, Fitzallen, & Chick, 2020). Similarly, in the design process, the role of engineering (e.g., knowledge of design) within the STEAM disciplines is clear (English et al., 2020). Thus, by combining data modeling and design processes in STEAM lessons, teachers can provide a learning context that integrates knowledge from different STEAM subjects and domains and guides the learning process through interdisciplinary problem-finding and solving, which helps highlight the nature of each STEAM subject and domain, with a focus on mathematics and engineering. However, it is not clear how to embed the data modeling and design processes into STEAM education at the PST education level nor to what extent PSTs gain insight into the nature of STEAM subjects and domains.

Table 8.1 Correspondence between the data modeling and design processes.

Data modelling process (Lehrer & English, 2018)	Design process (English et al., 2017)
DMP(a): Posing questions	DP(a): Problem framing or scoping
DMP(b): Generating, selecting and measuring attributes	DP(b): Generating multiple ideas; DP(c): Considering benefits and trade-offs
DMP(c): Collecting samples	N/A
DMP(d): Organizing, structuring, measuring, and representing data	N/A
N/A	DP(d): Designing and constructing
DMP(e): Making inferences	N/A
DMP(a): Reposing questions	DP(e): Testing; DP(f): Redesigning and reconstructing; DP(g): Reflecting and communicating

To help clarify these points, the following section demonstrates how STEAM education, incorporating the data modeling and design processes, can be implemented for Japanese PSTs.

PST education for STEAM activities combining data modeling and design processes

Setting

The participants were 15 mathematics PSTs (aged 20–21) in their third year at a Japanese university, who were studying to obtain teaching licenses for middle and high school (G7–G12) mathematics. The STEAM class was delivered as part of a mathematics research seminar, which constituted a specialized mathematics course required for these licenses. The seminar theme was "Mathematics and STEAM Education." The first author was the class teacher, and the PSTs worked in six groups of two to three individuals. The class consisted of 13 lessons (90 minutes each) divided into two units to develop generic competencies for interdisciplinary problem-solving. In the first seven-lesson unit, the teacher outlined the background of STEAM education in Japan and asked PSTs to work in groups on problems involving statistics (i.e., the paper helicopter experiment and the relationship between ice cream sales and weather) using the free data analytics education software Common Online Data Analysis Platform (CODAP). In the second six-lesson unit, the PSTs worked in groups on the *litter bin redesign task* (Figure 8.1). This section focuses on the latter task.

The litter bin redesign task is an authentic real-world problem that involves data-driven redesign of university litter bins for combustible waste. As it is difficult to create litter bins with the same dimensions, PSTs need to make prototypes/scale models of them using materials such as cardboard or origami paper (Figure 8.2). The development of these prototypes corresponds to the

Your university has litter bins for combustible waste, as shown in the picture. However, there is scope for improvement in the design of the litter bins in terms of their capacity, size, shape, environment, and sorting. Redesign the university litter bin based on the data to achieve an ideal design.

Figure 8.1 Litter bin redesign task.

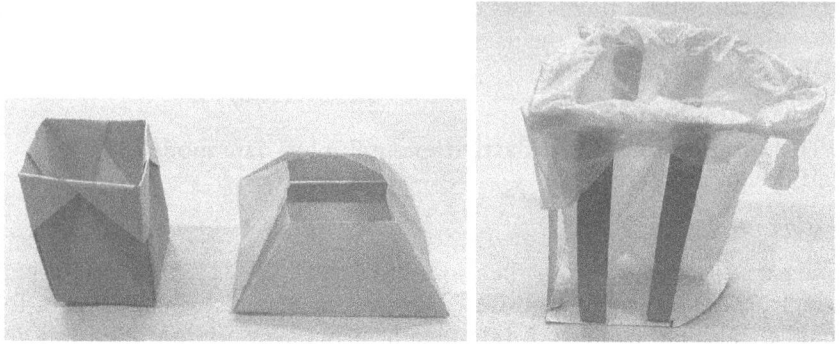

Figure 8.2 Prototypes/scale models of the litter bin.

engineering design process. Creating a prototype also involves collecting data on actual litter bins, analyzing the functional and statistical relationships between bin volume and the number of litter bins using CODAP, and predicting and estimating the optimal volume of litter bins and the quantity of litter that can be accommodated. Such data-based exploration constitutes a data modeling process. Traditionally, the task of modeling the relationship between volume and box shape has been included in junior and senior high school math textbooks in Japan. For example, the container-making task in Figure 8.3 was included in a **G9** mathematics textbook nearly 80 years ago in Japan (Chūtō-Gakko Kyōkasho Kabushiki Gaisha, 1943). In this task, PSTs were required to determine the length of one side of a corner square (x cm) and the volume of the container as a cubic function when the volume was 1 L. In addition to mathematical content, the litter bin redesign task also includes scientific (environmental) content, such as the relationship between the quantity of rubbish, carbon dioxide emissions, and rubbish sorting. It also encompasses liberal arts (humanities) content such as the economic benefits of installing litter bins and

一邊ノ長サガ 24 cm ノ正方形ノ厚紙ガア ル。ソノ四隅カラ同ジ大キサノ正方形ヲ切 落シテ,容積 1 *l* ノ箱ヲ作リタイ。

ドレダケノ大キサノ正方形ヲ切落シタラ ヨイカヲシラベテミョウ.

x cm

24 cm

From a square in which the one side length is 24 cm, we will make a container so that the volume becomes 1L by cutting squares of a similar size at the four corners of the container. Determine how big a square you need to cut off.

Figure 8.3 The container-making task in a Japanese G9 mathematics textbook from nearly 80 years ago (Chūtō-Gakko Kyōkasho Kabushiki Gaisha, 1943, p. 52).

the ethical aspects of sorting rubbish. Thus, the litter bin redesign task fulfils the requirements to be considered a STEAM task that also involves the data modeling and design processes.

The focus of the lessons was on PST group activities. The teacher's role in the classroom practice was to explain the task, observe the groups and offer advice if they were stuck or in trouble. In addition, each PST brought their laptop to class to process data in CODAP and Excel, and collect information from the Internet. The teacher provided origami paper, cardboard, tape, tape measures, scissors, marbles, and digital scales. At the end of each lesson, the PSTs wrote a report on their activities. At the end of the second unit, each group gave a final presentation on the design and justified the most suitable litter bin by using PowerPoint slides and their prototypes.

Implementation

This section provides an overview of the implementation of the second unit. At the beginning of the unit, the teacher presented the litter bin redesign task using the themes of local and university rubbish management problems. When the task was first presented, the teacher did not specify the scope for design improvement or the use of data-based thinking (as described in Figure 8.3), but asked each group to consider the definition and criteria for the "optimal" litter bin. The following criteria were suggested by the groups: "large capacity but not too big or small"; "mouth of litter bin should be narrow, but a wide mouth is also important so throwing away litter is easy"; "color and shape should be designed so the litter bins can be used for sorting"; "low production costs for litter bins"; "noise emission"; "no odor leakage"; and "automatic opening and closing of the lid"; among others. Accordingly, the need for trade-offs (English, 2020) was already apparent.

As the seminar theme was "Mathematics and STEAM Education," the teacher proposed initially focusing on litter bin capacity, size, and shape, as these elements are related to mathematics. The PSTs also recognized that they needed to employ data-based thinking. Initially, some groups collected general data, using tape measures to measure the dimensions of the litter bins around the university and observing the surroundings and posters where the litter bins were located (Figure 8.4). Additionally, some groups shared and developed their ideas by sketching the litter bin design concept on a whiteboard or tablet screen.

Halfway through the unit, the groups created and revised prototypes/scale models of the litter bin using origami or cardboard. However, rather than basing these models on mathematical and/or statistical reasoning and using data-based procedures, many groups created objects according to vague design ideas; the use of data and mathematics was not evident in the designs. The teacher instructed each group to consider the best design based on the data. Accordingly, four groups posed and solved Fermi problems (e.g., Ärlebäck & Albarracín, 2019) to predict and estimate the capacity of the actual litter

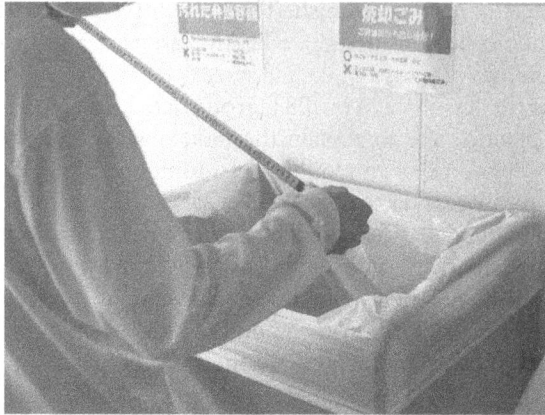

Figure 8.4 Measuring the dimensions of the actual litter bin.

bins to be subsequently applied to these prototypes/scale models. One group used marbles as rubbish to predict and estimate litter bin capacity, while others used origami paper to experiment and collect data on how rubbish enters the litter bin (Figure 8.5). Another group designed and conducted a survey

Figure 8.5 Experiment with model and origami to see how rubbish enters the bin.

Table 8.2 Main PST activities and the corresponding data modeling/design process phases.

Main PST activities	Corresponding phase(s)
Consideration of the definition and criteria for the "optimum" litter bin	DMP(a)(b); DP(a)(b)(c)
Examination of the dimensions, design, and function of actual litter bins and their sorting labels	DMP(b)(c); DP(b)
Creation of design sketch models of litter bins	DP(d)
Production and revision of litter bin prototypes	DMP(a); DP(d)(e)(f)(g)
Experimentation with prototypes/scale models	DMP(c)
Litter bin survey planning and execution	DMP(c)
Organization and analysis of collected data	DMP(d)
Estimation of litter bin capacity	DMP(e)
Sought optimal values by graphing the relationship between the two variables	DMP(e)

on litter bins among university students, using Excel and CODAP to visualize and analyze the data. Only one group obtained optimal values by graphing the relationship between the two variables (i.e. angle and size of litter bin opening) using CODAP.

To summarize the information above, Table 8.2 presents the main PST activities corresponding to each phase of the data modeling/design processes. This result shows the *process* integration in STEAM education. Table 8.3 lists the STEAM elements inherent in the PST activities. This result reveals the *content* integration in STEAM education.

Table 8.3 STEAM elements inherent in main PST activities.

Science	• Variable control
	• Controlled experiments
	• Considering pollution problems
Technology	• Use of Excel and CODAP
Engineering	• Creation of prototypes/scale models of litter bins
	• Optimization of litter bin size, capacity, shape, etc./imagining a realistic litter bin in terms of functionality, ease of installation, cost, etc.
	• Considering trade-offs in optimal litter bin design
Liberal arts	• Devising illustrations for waste-sorting
	• Considering universal design
	• Considering university budget
Mathematics	• Modeling
	• Use of ratios
	• Measurement
	• Fermi estimation
	• Organization, visualization, and analysis of data
	• Variation considerations
	• Logical explanation based on evidence

Exemplary design outcomes from one group

We summarize exemplary design results from a group of complementary data modeling and design processes. The group designed the best litter bin in terms of ease of sorting (i.e., clearly marked to facilitate waste sorting), ease of disposal (i.e., a large disposal opening), and optimal capacity (i.e., not too large or small).

Regarding ease of sorting, the group prepared and conducted a questionnaire with university students ($n = 207$) asking whether they were aware of waste-sorting in general, how much they knew about the sorting required by the university, and what they throw away in the university's litter bins for combustibles. They then collated the responses. The results indicated that many students felt that although they were aware of waste-sorting, there were too many types of sorting and it was difficult to know what type of waste to put in which litter bin. Students also threw away relatively small items such as tissues in litter bins. Using these results, the group determined that, from a universal design perspective, both text and illustrations should be placed on litter bins to help communicate waste-sorting.

In terms of ease of disposal, the group designed two prototypes/scale models of litter bins with different opening shapes, but the same height (Figure 8.6). They created origami rubbish models (five different sizes), conducted an experiment in which garbage was thrown in each prototype/scale model 50 times (Figure 8.5), and compared the percentage of rubbish that entered the litter bin (collection rate). They considered how the variability and different conditions of the experimental method (i.e., throwing angle and distance) influenced the experimental results. They found that the collection rate of Litter Bin A was 60% and that of Litter Bin B was 94%; therefore, the group adopted the opening shape of Litter Bin B.

Finally, regarding ideal capacity, the group determined the annual quantity of combustible waste collected at their university and the number of litter bins on campus (Figure 8.1) from university reports, and estimated that

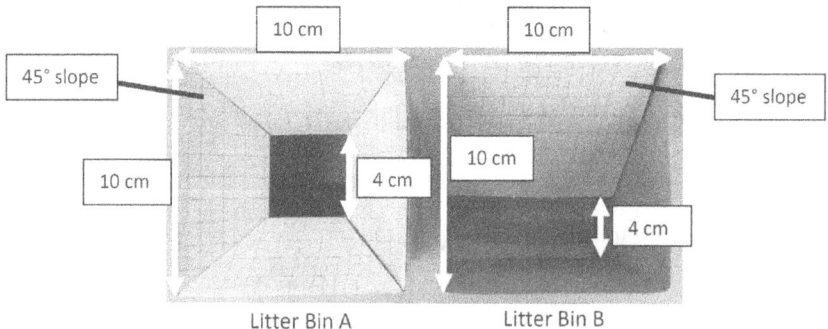

Figure 8.6 Two litter bin prototypes/scale models with different opening shapes.

$$V = (\tfrac{4}{3} \times 3.14 \times 2^3) \times 3000$$

Volume per tissue   Number of tissues for 3 kg

Weight

$$= 100,480cm^3 \doteqdot 100,000cm^3 \ (100L)$$

Volume of waste per week (Monday to Friday) I 1週間のゴミの体積

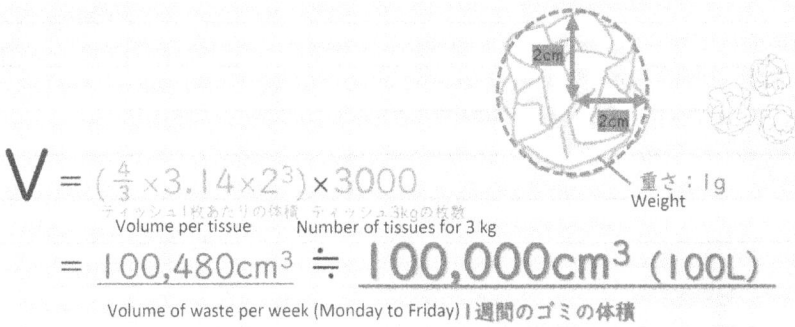

Figure 8.7 The group's presentation slide on rubbish modeling and estimation of the amount of rubbish in the litter bins.

approximately 600 g of rubbish was generated per litter bin per day. From this calculation, they estimated that approximately 3 kg of waste was generated per litter bin from Monday to Friday. Based on the results of the questionnaire survey, the group then assumed that the most rubbish in the litter bins comprised tissues, and subsequently modeled tissues as a sphere with a weight of 1 g and a radius of 2 cm (Figure 8.7). The amount of rubbish generated in the litter bins from Monday to Friday was estimated as shown in Figure 8.7, and an optimal litter bin was designed in terms of both ideal volume (Figure 8.8) and universal design (Figure 8.9).

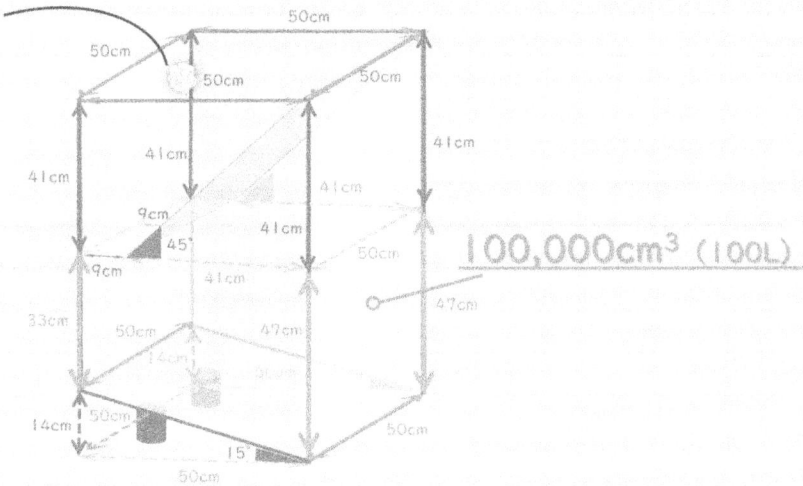

$$100,000cm^3 \ (100L)$$

Figure 8.8 The group's presentation slide on the litter bin designed in terms of ideal volume.

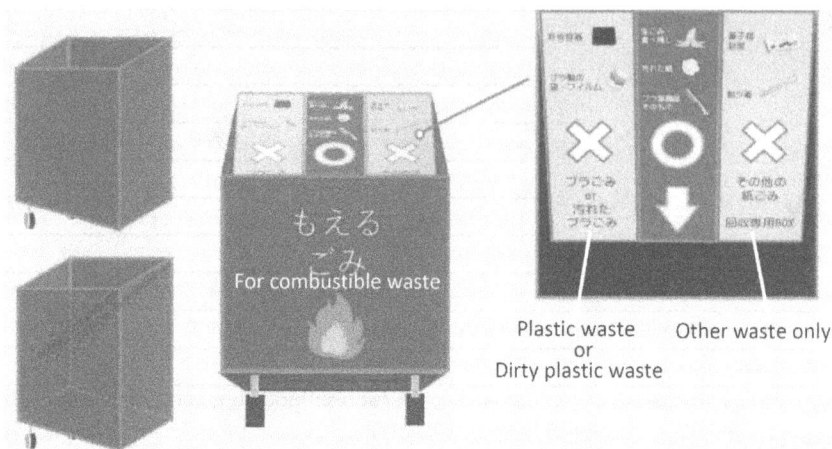

Figure 8.9 The group's presentation slide on the litter bin designed in terms of universal design.

PST perceptions of the roles of mathematics and mathematics teachers in STEAM education

At the end of the second unit, the teacher asked the PSTs to reflect on the roles of mathematics and mathematics teachers in STEAM education and summarize them in a report. A PST from the abovementioned group reflected on the STEAM practice activities (e.g., modeling in designing as shown in Figure 8.7), and realized that mathematics played a role in facilitating data-based decision-making, and providing models for concretization and abstraction in interdisciplinary problem-solving:

> … I believe that mathematics in STEAM education helps students use data correctly, make optimal decisions, and derive correct results, as they learn to incorporate elements of mathematics and using data in other subjects … I believe that it also helps students cultivate mathematical thinking of concretization and abstraction. For example, it is possible to promote more flexible thinking by concretizing abstract things or using abstraction to model concrete things rather than taking everything as it really is. In fact, when we calculated the volume of the litter bin, we were able to derive an approximate reasonable value by considering the tissue as a sphere and assuming that it would fill the volume of the litter bin [as shown in Figure 8.7] …

In the same report, the PST reflected on the role of mathematics teachers in STEAM education, and referred to teachers as designers of instructional tasks (Kaur et al., 2022), who make students aware of the relationship between mathematics and other subjects. It demonstrates that the PST's perspective of mathematics education improved, and that this change was due to the activities in the STEAM practice:

> … In fact, I realized that mathematics was very useful in other non-mathematics situations through the STEAM activities. Therefore, I believe that mathematics teachers need to set up situations (tasks) in which it is easy for students to feel connected with other subjects. In the mathematics classes I have attended so far in my schooling, letters and equations appeared out of the blue, the examples were all mathematical events (idealized contents), and I often had to solve difficult problems that clearly reflected characteristic of the subject "pure mathematics" …

Discussion and implications for STEAM PST education

This chapter explored how STEAM education incorporating data modeling and design processes could be taught to Japanese PSTs. Tables 8.2 and 8.3, and the group case study demonstrated a STEAM education practice involving the design of an optimal litter bin by applying data modeling and design processes; linking knowledge and practice in mathematics, science, and the arts; and using technology where appropriate. The integration of both process and content in STEAM education was evident in the PSTs' activities. In particular, the data-based activities required PSTs to logically/empirically present objective evidence and use mathematics; while, the design-based activities required them to flexibility create prototypes and scale models, which naturally resulted in modeling and model-based experimentation. In addition, liberal arts topics, such as universal design and design economics were incorporated into these activities. In this way, the data modeling and design processes enabled STEAM education in terms of process and content.

Both data modeling and design processes are universal processes common to STEAM subjects and domains (English, 2020; Hjalmarson et al., 2020; Lehrer & English, 2018). By combining the data, modeling and design elements of these processes, the interconnectedness of STEAM subjects and domains with a focus on processes and contents in mathematics and engineering becomes more apparent. In particular, PSTs' reflections during the STEAM lessons revealed that the modeling process in designing was key to elevating the role of mathematics in STEAM education beyond lending calculation skills. While the current study focused on pre-service mathematics teachers, the same method could be applied to PST education relating to other STEAM subjects, due to the interdisciplinary nature of the data modeling and design processes. PSTs specializing in other subjects could reflect on the role of the subject in relation to data, modeling, and design. Thus, one strategy for implementing STEAM education that highlights the role of each subject and domain is to include activities that incorporate data modeling and design processes.

In practice, however, three issues have emerged from the current study with regard to deepening the quality of integration. First, the level of mathematics used by the PSTs, and the level of cognitive demand required by the mathematical tasks in the lessons corresponded to primary and lower secondary schools, while the technology was limited to data visualization and simple analysis. This result confirms the ever-present challenge of increasing the level of mathematics within STEAM activities (Forde et al., 2023). Second, the current study exposed a lack of STEAM experience among PSTs, which was also pointed in the previous studies (e.g., Enderson et al., 2020). Therefore, in the future, it is essential to develop a curriculum and system for STEAM education at the PST education stage. Third, the current study highlighted the need for collaboration among PSTs and teacher educators of STEAM subjects and disciplines (Isozaki & Isozaki, 2021), as the litter bin redesign task requires a wide range of interdisciplinary knowledge. Although this study is based on a Japanese sample, we believe that the importance and issues of STEAM PST education emphasized are applicable to education in other countries. Therefore, it is important that STEAM PST education is implemented in different countries in the future, and that the results and strategies for improvement are shared.

Acknowledgment

This study was supported by the Japan Society for the Promotion of Science KAKENHI Grant Numbers JP20K20832, JP21K02513, JP20H01739, JP21H00191 (JP23K20744).

References

Anderson, J., English, L., Fitzallen, N., & Symons, D. (2020). The contribution of mathematics education researchers to the current STEM education agenda. In J. Way, C. Attard, J. Anderson, J. Bobis, H. McMaster, & K. Cartwright (Eds.), *Research in mathematics education in Australasia 2016–2019* (pp. 27–57). Springer.

Ärlebäck, J. B., & Albarracín, L. (2019). The use and potential of Fermi problems in the STEM disciplines to support the development of twenty-first century competencies. *ZDM-Mathematics Education, 51*, 979–990. https://doi.org/10.1007/s11858-019-01075-3

Chūtō-Gakko Kyōkasho Kabushiki Gaisha (1943). *Suugaku (Tyugakkoyou) 3 Dai Ichirui (Mathematics (for secondary school) 3 category1)*. Okura Insatsujyo [in Japanese].

Cobb, G. W., & Moore, D. S. (1997). Mathematics, statistics, and teaching. *The American Mathematical Monthly, 104*, 801–823. https://doi.org/10.2307/2975286

Drake, S. M., & Burns, R. C. (2004). *Meeting standards through integrated curriculum*. Association for Supervision and Curriculum Development.

Enderson, M. C., Reed, P. A., & Grant, M. R. (2020). Secondary STEM teacher education. In C. C. Johnson, M. J. Mohr-Schroeder, T. J., Moore, & L. D. English (Eds.), *Handbook of research on STEM education* (pp. 349–360). Routledge.

English, L. (2017). Advancing elementary and middle school STEM education. *International Journal of Science and Mathematics Education, 15*, 5–24. https://doi.org/10.1007/s10763-017-9802-x

English, L. (2020). Facilitating STEM integration through design. In J. Anderson & Y. Li (Eds.), *Integrated approaches to STEM education* (pp. 45–66). Springer.

English, L., Adams, R., & King, D. (2020). Design learning in STEM education. In C.C. Johnson, M. J. Mohr-Schroeder, T. J. Moore, T. J., & L. D. English (Eds.), *Handbook of research on STEM education* (pp. 76–86). Routledge.

English, L., King, D., & Smeed, J. (2017). Advancing integrated STEM learning through engineering design: Sixth-grade students' design and construction of earthquake resistant buildings. *The Journal of Educational Research, 110,* 255–271. https://doi.org/10.1080/00220671.2016.1264053

Fan, S., & Yu, K. (2017). How an integrative STEM curriculum can benefit students in engineering design practices. *International Journal of Technology and Design Education, 27,* 107–129. https://doi.org/10.1007/s10798-015-9328-x

Forde, E. N., Robinson, L., Ellis, J. A., & Dare, E. A. (2023). Investigating the presence of mathematics and the levels of cognitively demanding mathematical tasks in integrated STEM units. *Disciplinary and Interdisciplinary Science Education Research, 5*(3), 1–18. https://doi.org/10.1186/s43031-022-00070-1

Hestenes, D. (2010). Modeling theory for math and science education. In R. Lesh, P. Galbraith, C. Haines, & A. Hurford (Eds.), *Modeling students' mathematical modeling competencies* (pp. 13–41). Springer.

Hjalmarson, M., Holincheck, N., Baker, C. K., & Galanti, T. M. (2020). Learning models and modelling across the STEM disciplines. In C. C. Johnson, M. J. Mohr-Schroeder, T. J., Moore, & L. D. English (Eds.), *Handbook of research on STEM education* (pp. 223–233). Routledge.

Isozaki, T., & Isozaki, T. (2021). Nihon-gata STEM kyōiku no kouchiku ni mukete no rironteki-kenkyū–Hikaku-kyōikugakuteki-shiza kara no bunseki wo tōshite (Theoretical research for establishing a Japanese-style STEM education: Analysis for a comparative historical point of view). *Journal of Science Education in Japan, 45*(2), 142–154 [in Japanese]. https://doi.org/10.14935/jssej.45.142

Kaur, B., Leong, Y. H., & Attard, C. (2022). Teachers as designers of instructional tasks. *Mathematics Education Research Journal, 34,* 483–489. https://doi.org/10.1007/s13394-022-00437-7

Kawakami, T. (2017). Combining models related to data distribution through productive experimentation. In G. A. Stillman, W. Blum, & G. Kaiser (Eds.), *Mathematical modelling and applications: Crossing and researching boundaries in mathematics education* (pp. 95–105). Springer.

Kawakami, T. (2022). Teigakunen-zidō no hi-keisikiteki-na-tōkeiteki-suisoku no sokusin ni okeru moderu no yakuwari: dēta-moderingu no katei ni tyakumoku site (The role of models in promoting informal statistical inferences of lower grade children: Focusing on data modeling processes). *Journal of Science Education in Japan, 46*(2), 125–140 [in Japanese]. https://doi.org/10.14935/jssej.46.125

Kawakami, T., & Saeki, A. (2021). Sansū-sūgaku kyōkasyo no kyōzai kara STEM-kyōzai heno sai-kyōzaika: Sūgaku-kyōiku no tatiba karano STEM-kyōsi-kyōiku heno iti-teian (Retransformation of math-textbook tasks into integrated STEM tasks: A proposal to STEM teacher education from mathematics education perspective). *JSSE Research Report, 35*(5), 79–84 [in Japanese]. https://doi.org/10.14935/jsser.35.5_79

Koehler, M. J., Mishra, P., & Yahya, K. (2007). Tracing the development of teacher knowledge in a design seminar: Integrating content, pedagogy, and technology. *Computers and Education, 49,* 740–762. https://doi.org/10.1016/j.compedu.2005.11.012

Lehrer, R., & English, L. (2018). Introducing children to modeling variability. In D. Ben-Zvi, K. Makar, & J. Garfield (Eds.), *International handbook of research in statistics education* (pp. 229–259). Springer.

Lesh, R. A., & Doerr, H. M. (Eds.). (2003). *Beyond constructivism: Models and modeling perspectives on mathematics problem solving, learning, and teaching*. Routledge.

Maass, K., Geiger, V., Ariza, M. R., & Goos, M. (2019). The role of mathematics in interdisciplinary STEM education. *ZDM-Mathematics Education, 51*, 869–884. https://doi.org/10.1007/s11858-019-01100-5

Ministry of Education, Culture, Sports, Science and Technology (MEXT) (Central Council for Education) (2021). *"Reiwa no nihongata-gakkōkyōiku" no kouchiku wo mezashite: Subete no kodomotachi no kanousei wo hikidasu, kobetusaitekinamnabi to kyōdoutekina manabi no jitugen (toushin) (Toward the construction of "Japanese style school education": Optimal individualized learning and collaborative learning that bring out the potential all children (report))* [in Japanese]. Retrieved from https://www.mext.go.jp/content/20210126-mxt_syoto02-000012321_2-4.pdf

Moore, T. J., Johnston, A. C., & Glancy, A. W. (2020). STEM integration: A synthesis of conceptual frameworks and definitions. In C. C. Johnson, M. J. Mohr-Schroeder, T. J. Moore, T. J., & L. D. English (Eds.), *Handbook of research on STEM education* (pp. 3–16). Routledge.

Nishimura, K., & Tachikawa, S. (2019). STEM-kyōiku ni muketa sūgaku-ka no kyōsikyōiku ni kansuru iti-kōsatu: EU no Mascil project no bunseki wo tōsite (A study on professional development for mathematics teachers on STEM education: Through analyzing the Mascil project in EU). In *Proceedings of the 43rd Annual Meeting of Japan Society for Science Education* (pp. 151–154). JSSE [in Japanese]. https://doi.org/10.14935/jssep.43.0_151

Petrosino, A. J. (2016). Teachers' use of data, measurement, and data modelling in quantitative reasoning. In R. Duschl & A. S. Bismack (Eds.), *Reconceptualizing STEM education* (pp. 167–180). Routledge.

Smith, C., Fitzallen, N., Watson, J., & Wright, S. (2019). The practice of statistics for stem: Primary students and pre-service primary teachers exploring variation in seed dispersal. *Teaching Science, 65*(1), 38–47.

Tytler, R., Williams, G., Hobbs, L., & Anderson, J. (2019). Challenges and opportunities for a STEM interdisciplinary agenda. In B. Doig, J. Williams, D. Swanson, R. Borromeo Ferri, & P. Drake (Eds.), *Interdisciplinary mathematics education: The state of the art and beyond* (pp. 51–81). Springer.

Watson, J., Fitzallen, N., & Chick, H. (2020). What is the role of statistics in integrating STEM education? In J. Anderson & Y. Li (Eds.), *Integrated approaches to STEM education* (pp. 91–115). Springer.

Watson, J., Fitzallen, N., English, L., & Wright, S. (2020). Introducing statistical variation in Year 3 in a STEM context: Manufacturing licorice. *International Journal of Mathematical Education in Science and Technology, 51*(3), 354–387. https://doi.org/10.1080/0020739X.2018.1562117

Yata, C., Ohtani, T., & Isobe, M. (2020). Conceptual framework of STEM based on Japanese subject principles. *International Journal STEM Education, 7*(12), 1–10. https://doi.org/10.1186/s40594-020-00205-8

9 Scientific inquiry required in STEAM education movement

Ryugo Oshima

School subjects related to STEAM education

Japan envisions Society 5.0 as "a human-centered society that balances economic advancement with the resolution of social problems by a system that highly integrates cyberspace and physical space" (Cabinet Office, n.d.). This vision anticipates a drastic transformation in our lifestyle and economy that should ultimately benefit people. However, Japan faces challenges in its transition to this digital society, including "a lack of research and educational opportunities in mathematics and information science and a shortage of human resources with high-level knowledge and skills in information science" (Ministry of Education, Culture, Sports, Science and Technology [MEXT], 2018, p. 4).

In recognition of these critical future-oriented issues, three key competencies have been proposed:

> the ability to interpret sentences and information accurately and engage in dialogue, the ability to think and examine scientifically and apply such knowledge, and the ability and the sensibility to identify values and the curiosity and the ability to inquire.
>
> (MEXT, 2018, p. 7)

The directions of educational policy have been set according to the school stages, with STEAM education being expected to be conducted at the upper secondary school and university levels. Two subjects at the upper secondary school level have been identified as core to STEAM education because they align with the common aims of STEAM: "setting events in the complex contexts of real life and society as cross-curricular topics and developing an inquiry process by integrally applying what has been studied in each traditional subject to problem solving" (MEXT (Central Council for Education), 2021, p. 57).

Contrastingly, sound academic abilities are anticipated to be nurtured at the elementary and lower secondary school levels in preparation for Society 5.0

DOI: 10.4324/9781003392545-10

(MEXT (Central Council for Education), 2021). Developing sound academic abilities should primarily equate to acquiring knowledge and skills in each subject. In this STEAM education model, students are expected mainly to acquire knowledge and skills in each subject at the elementary and lower secondary school levels and then apply them in STEAM education at the higher levels. The critical point is whether the competencies developed in the process of acquiring knowledge and skills in each subject are sufficient for their application in STEAM education. To cultivate such competencies in each subject by the end of lower secondary school, we need to identify the competencies necessary for STEAM education at the higher secondary school level.

Period for inquiry-based cross-disciplinary study

From elementary to upper secondary school, period for integrated studies has been implemented. This period is considered an essential time for schools to conduct cross-disciplinary and integrated studies beyond the confines of traditional subjects, according to the realities of the region, school, and students. It is also considered an important time to promote inquiry-based and collaborative learning. However, there has been variance among schools regarding what qualities and abilities should be nurtured through period for integrated studies and how it relates to other subjects.

Therefore, it was deemed necessary to clarify the positioning of period for integrated studies from a perspective that emphasizes more inquiry-based activities in upper secondary schools (MEXT, 2023). In upper secondary schools, the name was changed to period for inquiry-based cross-disciplinary study. Building on the efforts of period for integrated studies in elementary and lower secondary schools, the aim is now to foster students' ability to find and explore questions by combining and integrating "perspectives and ways of thinking" in line with the characteristics of each subject and relating them to students' career direction, in light of their own way of being and living. Although schools have discretion over the implementation year and number of hours for period for inquiry-based cross-disciplinary study, it is expected that when it is implemented across all three years, approximately one to two hours per week should be secured.

The differences from inquiry in each subject, such as science, have been indicated in three points (MEXT, 2023, p. 10). The first point is that the area of study in this subject do not remain within a particular subject; it is cross-disciplinary and comprehensive. Period for inquiry-based cross-disciplinary study targets events that exist in the complex context of real society and real life. The second point is inquiry by integrally applying perspectives and ways of thinking in multiple subjects. Whereas other inquiries aim to deepen the understanding in each subject, period for inquiry-based cross-disciplinary study requires students to examine and find out problems that exist in the complex context of real society and real life from various angles. The third point is that learning activities in this subject emphasize finding optimal or satisfactory solutions to problems where the path to resolution is not immediately clear or where there is no single correct answer.

The contemplation of one's own way of being and living is considered from the following three perspectives (MEXT, 2023, p. 14). The first perspective is thinking about what one should do and how one should act as a member of society and nature, and as a human being, in relation to people, society, and nature. The second perspective is thinking about the meaning and value of learning for oneself. The third perspective is thinking about how what one has learned is connected to one's current and future way of being and living. In this way, period for inquiry-based cross-disciplinary study is also expected to play a role in career education.

Inquiry topics to be addressed are divided into four categories, as shown in Table 9.1, where further subcategories are indicated (MEXT, 2023, pp. 87–89). All of the categories go beyond the scope of existing subjects and are comprehensive in content. In addition, there is a category directly related to career education.

In this way, in period for inquiry-based cross-disciplinary study, inquiry is emphasized and activities are carried out to address issues in real society and real life using the perspectives and ways of thinking cultivated in each subject the students have studied thus far. Students tackle these issues in their own way. The aim is to shape, through these activities, the direction of students' careers.

Inquiry-based study of science and mathematics

Inquiry-based study of science and mathematics is organized into basic inquiry-based study of science and mathematics and inquiry-based study of science and mathematics. These subjects can replace part or all of period for inquiry-based cross-disciplinary study. The year of implementation and number of hours are at the discretion of each school; however, if each subject is to be taken in a single year, approximately one hour per week is required for basic inquiry-based study of science and mathematics and approximately two

Table 9.1 Inquiry topics in period for inquiry-based cross-disciplinary study.

Cross-disciplinary and comprehensive issues in response to contemporary challenges
International understanding, information, environment, welfare, and health
Issues according to the characteristics of the region and the school
Town development, traditional culture, regional economy, disaster prevention, urban planning, tourism
Issues based on students' interests
Creation of culture, education and childcare, life and medical care
Issues related to occupations and students' future paths
Occupation, labor

to five hours per week for inquiry-based study of science and mathematics. Basic inquiry-based study of science and mathematics and inquiry-based study of science and mathematics do not need to be taken in a specific order; however, they are designed with a phased set of objectives and content, and it is desirable for students to take inquiry-based study of science and mathematics after completing basic inquiry-based study of science and mathematics (MEXT, 2019, p. 15). However, if the qualities and abilities aimed to be nurtured in basic inquiry-based study of science and mathematics have already been developed in period for inquiry-based cross-disciplinary study or other subjects, inquiry-based study of science and mathematics can be taken without basic inquiry-based study of science and mathematics being taken first.

Inquiry-based study of science and mathematics aims to foster "knowledge and skills necessary for inquiry," "the ability to grasp issues from multiple and complex perspectives, to set and investigate problems, and to solve them," and "the attitude to face various events and challenges, to think and act persistently, and to strive to solve problems" (MEXT, 2019, p. 29). It is a subject that promotes inquiry by capturing phenomena from diverse and complex perspectives without being confined to the framework of subjects and by utilizing or combining mathematical and scientific perspectives with rich ideas. The goal is for students to investigate more actively and challengingly. In addition, rather than the focus being the presence or value of new insights as the result of the inquiry, focus is placed on the student's thinking and attitude during the inquiry process and on fostering the competency necessary to actively complete the entire inquiry process (MEXT, 2019, p. 16).

In basic inquiry-based study of science and mathematics, students investigate one or more issues from "natural phenomena and social phenomena," "cutting-edge science and interdisciplinary fields," "natural environment," "science and technology," and "mathematical phenomena" and are made to create reports and presentations (MEXT, 2019, p. 24). Although the subject name includes "science and mathematics," the integration of science and mathematics is inferred to not be a mandatory element.

As reflected in the above descriptions, inquiry-based study of science and mathematics places more emphasis on students' inquiry-based activities than existing subjects such as science. However, the content handled does not necessarily have to transcend academic disciplines. In inquiry-based study of science and mathematics, more emphasis is placed on conducting inquiry-based activities or fostering inquiry-based abilities than on integrating content.

STEAM library as a platform for STEAM education

The role of STEAM library

The Ministry of Economy, Trade and Industry (METI) of Japan asserts that STEAM education should not only be conducted during period for integrated studies or period for inquiry-based cross-disciplinary study but should also extend to other subjects and be implemented across disciplines (METI, 2020). However, in promoting STEAM education, they identified challenges such as a lack of STEAM learning programs that guide students, each with different interests and concerns, to inquiry, and a lack of established models and evaluation methods to realize cross-disciplinary and integrated class organization using such programs. To overcome these challenges, they developed a platform called the "STEAM Library," where teaching materials are provided. This initiative suggests that, in STEAM education at METI, cross-disciplinary and inquiry-based activities are emphasized in the rationale.

Learning components in the STEAM library

As of the end of 2021, the STEAM Library contained 67 components. Each component showed a relationship with subjects and the Sustainable Development Goals (SDGs) and was composed of multiple learning activities. The following are three characteristics of these components (Masumoto, 2022).

First, regarding the relationship between the components and the subjects, of the 67 components, 65 were related to multiple subjects (including learning activities related only to period for integrated studies). The two components that were only positioned in one subject were both related to information science and the titles of the components were "AI Human Resource Development Course (Practical Course)—How is AI used in society?" and "AI Human Resource Development Course (Theoretical Course)—How is AI made?" If we define the science subjects as science, information, arithmetic/mathematics, and technology, the number of components related to each subject was 44, 39, 31, and 10, respectively. However, if we define social studies, foreign languages, the arts, Japanese language, home economics, and physical and health education—that is, liberal arts subjects—as the "A" in STEAM, the number of components related to each subject was 49, 28, 28, 27, 16, and 11, respectively. Thus, the majority of the components in the STEAM Library are cross-curricular and are widely related not only to science subjects but also to liberal arts subjects. In particular, there are many components related to science, information, and arithmetic/mathematics in the science subjects and substantially more components related to social studies in the liberal arts subjects.

Second, regarding the relationship with the SDGs, among the 17 goals of the SDGs, those that had the most related components were "9: Build resilient infrastructure, promote sustainable industrialization and foster innovation," "8: Promote inclusive and sustainable economic growth, employment and decent work for all," and "3: Ensure healthy lives and promote well-being for all at all ages," with 40, 33, and 32 related components, respectively. Most of the components of the STEAM Library are provided by companies. Therefore, there were numerous components related to SDGs in industry and the economy. In fact, issues faced by the companies in real society and real life were mainly dealt with. Thus, many of the components have a strong element of career education.

Third, regarding the relationship with inquiry-based learning, period for inquiry-based cross-disciplinary study requires learning through four processes of problem-solving: (1) setting the problem; (2) collecting information; (3) organizing and analyzing; and (4) summarizing and expressing (MEXT, 2023, p. 6). Therefore, we examined whether the components were in a format that could provide inquiry-based learning in these four processes. As a result, worksheets were used in many components and the structure enabled the class to easily proceed along the four processes of inquiry. However, it was not structured for students to proceed with inquiry-based learning on their own, such as by providing problems or information.

Current state of science teaching and learning in Japan

Lacorte et al. (2022) described several characteristics of science education in a Japanese lower secondary school, as observed by a foreign visitor (a Filipino science teacher who is also a co-author). The teacher attended the school every Monday and Tuesday for five months, from 7:50 a.m. to 4:00 p.m., observing not only science classes but the entire day's school program. In this section, we will examine the characteristics of representative Japanese science lessons as observed by the teacher and discuss the challenges that lower secondary science lessons in Japan may present when transitioning to STEAM education in upper secondary schools.

Science lesson practices in lower secondary school

The study by Lacorte et al. (2022) revealed a number of findings. One of the characteristics of science classes in Japan that was observed by the Filipino teacher is the frequent opportunities for students to conduct experiments and make observations. Conversely, lecture-style classes composed solely of teacher-led instruction are not the norm in Japanese science classrooms. For science lessons that include laboratory work and observations, the teacher typically begins by writing the experimental procedures on the board and discussing with the students what they will be doing. The students then copy these procedures into their notebooks. During the experiment or observation, the students independently retrieve the necessary equipment from the lab shelves and cabinets and set up the experiment. The teacher checks and assists with their preparations, then moves around the lab from group to group to monitor and support the students' work. After the experiment or observation, the students discuss the results within their groups, then return their equipment and clean their workspace. In the following lesson, the students discuss their results with the whole class. After the discussion, the teacher provides a summary of the lesson.

According to the observer's account in Lacorte et al. (2022), an inquiry-based science lesson generally includes the following steps: (1) stimulating curiosity by posing questions or presenting interesting problems to the class; (2) allowing the students to think about these questions or problems and then asking them to propose hypotheses or tentative solutions; (3) encouraging the students to discuss their answers with their classmates (usually in groups of 3 or 4); (4) conducting experiments or investigations to test their initial answers or hypotheses; (5) observing and collecting data; (6) communicating the results to the entire class; and (7) comparing their results with others' answers after the teacher summarizes the lesson.

In this inquiry-based learning process, the teacher's role remains as a primary source of information and the students' activities are still largely guided by the teacher. Oshima et al. (2017) also indicated that Japanese teachers tend to maintain communication even during experiments. Although many students perceive that they are given the opportunity to make inferences, formulate ideas based on experimental data, explain their own ideas, and propose hypotheses, they also often feel that they are required to follow the experiment's procedures (Lacorte et al., 2022). Students appear to enjoy their limited freedom to engage in scientific processes while adhering to the teacher's strict instructions. Although students do not have absolute autonomy in conducting scientific inquiries in class, they appear content with their limited freedom.

How students work in science

Students have reported being interested in science lessons, finding them easy to understand and actively participating in them (Lacorte et al., 2022). In addition, observations suggest they have a strong foundation in self-management, especially in terms of discipline and control during lessons. Indeed, Sageme and Oshima (2023) reported that approximately 90% of students ($n = 205$) responded positively to the question "Do you work hard in science classes?" For the reasons behind students' diligence in science, approximately 30% of the students aimed for good grades, approximately 25% stated it was fun, approximately 25% sought a better understanding of science and the world, and approximately 10% admitted it was challenging. These results indicate that students not only find science enjoyable but are also motivated to understand the subject and achieve good grades. According to Lacorte et al. (2022), approximately 40% of surveyed students from Grade 7 to 9 ($n = 205$) supplement their school science studies with after-school learning platforms such as cram schools. However, most students perceive science careers as unexciting and have a low awareness of such careers, even though they acknowledge the potential rewards (Sageme & Oshima, 2023). When students were asked to rank their school subjects from most to least favorite, the results indicated a range of students' interest in science, with 9.3% ranking it as their top choice, 7.8% second, 19.3% third, 11.3% fourth, 10.3% fifth, 9.8% sixth, 7.4% seventh, 8.8% eighth, 10.3% ninth, 10.3% tenth, and 4.4% eleventh choice (Sageme & Oshima, 2023). Despite this variation in interest in science, the majority of students enjoy learning science and work diligently at it, both in school and through after-school platforms. However, this diligence is not necessarily because students aspire to pursue science careers in the future but because they find learning science fun, aim to earn good grades to secure their upper secondary school entry, and wish to understand the world and natural phenomena. Students are motivated to learn science through a combination of intrinsic and extrinsic motivations. On the other hand, Sageme and Oshima (2023) reported that students do not see relevance of school science to daily life so much and that they are unaware that science can be applied to other non-science subjects such as music. In this era of STEAM education, students are not just required to understand science itself but also to apply their scientific knowledge to solve real-world problems and comprehend the interconnectedness of scientific disciplines. From the students' perspectives, "the scope of science in lower secondary schools in Japan is too confined to specific scientific fields," as Tsuruoka (2017, p. 125) highlighted.

Lower secondary school science preparing for upper secondary school STEAM education

In science, it is critical not only to develop knowledge and skills but also to realize inquiry-based learning. Inquiry-based learning is important not only as preparation for full-fledged STEAM education in upper secondary schools but also for cultivating sound academic abilities in subjects. However, based on the challenges identified in the previous section and the students' perceptions, teachers strive to facilitate inquiry-based learning by involving students in the inquiry process, yet the students are passively focused on acquiring established "correct" scientific knowledge and skills. Scientific activities are human activities and should be more flexible. In considering the integration of scientific learning that connects to STEAM education in real society and real life, it is necessary to seek classes where students feel they can express their own thoughts and values more, encouraging an inquiry mind rather than idolizing scientific achievements. How can these two aspects, cultivating sound academic abilities and realizing inquiry-based learning, coexist in a classroom setting? In the following sections, we discuss two examples.

Fostering students' decision-making on procedures in a guided inquiry

Teachers pay close attention to how students think while learning specific content, tailoring science lessons to their individual learning situations. Although students can acquire scientific knowledge and skills through these processes, they often feel obligated to follow the procedures set forth by their teachers, resulting in what is often referred to as a recipe approach. Parkinson (2004) critiqued this method, arguing that students learn little as they mindlessly follow the steps without understanding why they're doing so.

The main reason for using the recipe approach is time constraint, as teachers must convey a large amount of scientific knowledge within a limited time.

However, this method does not adequately prepare students for "finding the optimal or satisfactory solution for problems that do not have a clear path to resolution or a single correct answer," (MEXT, 2023, p. 10) which is an expectation in period for inquiry-based cross-disciplinary study in upper secondary school STEAM education. This form of education aims to develop abilities to solve real-world problems and conduct authentic science research, where no absolute answer exists and students must make decisions independently. Thus, how can we incorporate opportunities for students' decision-making in a limited lesson time?

Bennett (2003) suggested that, since the 1980s, there has been a shift in England and Wales from a model emphasizing control of variables to a model stressing the importance of the evaluation of evidence. Instead of focusing on data collection, students are increasingly asked to examine experimental methods for interpreting data as evidence, considering the validity and reliability of the experimental methods.

In Oshima (2020), a lesson was implemented in 2015 in which students had to make decisions regarding the validity of experimental methods while planning their own experiments. In this study of Hooke's law, targeting 138 Grade 7 students separated into four class groups, the tasks and experimental apparatus were provided by the teacher. The elements that students had to decide on were mainly only the independent and dependent variables, frequency, interval, and range.

Interestingly, researchers and teachers had different perceptions of this lesson. Some researchers felt the activity might be too simple for the students. Conversely, the teachers at the school where the lesson was implemented thought it would be challenging for the students because they had never devised their own experimental plans before, speculating that most of the students would probably be unable to describe their experimental plans at all. In the end, most students managed to create their own experimental plans and carry out the experiment, indicating that the teachers' concern was misplaced. However, this outcome does not necessarily mean that the students' experimental plans were scientifically valid. The survey results of Oshima (2020) indicated that students actively engaged in the experimental activity and were satisfied. However, of the 117 students who provided free descriptions, approximately 80% described negative impressions such as finding the experimental planning difficult and feeling anxious. From further analysis, among the students with negative perceptions in their free descriptions, approximately 90% mentioned at least one positive item, suggesting that the researchers' concerns were also misplaced. These results indicate that the students had both positive and negative reactions to the activity. They felt satisfied with the activity, but they also expressed anxiety and found the task difficult. Additionally, based on the study results, most students believed that they recognized the purpose and procedure of the experiment and the importance of the experimental planning activity by devising their own experimental plan. Normally, in science experiments, teachers explain the purpose and method of the experiment to students and share it with them before performing the experiment. In the experimental activity of Oshima (2020), the Hooke's law experiment was simple; thus, a deep understanding of the purpose and method might appear necessary. However, students recognized that they understood the purpose and method of the experiment by planning the experiment themselves. Before planning the experiment, they might have understood the purpose and method in syntactic terms; however, they likely had not yet reached the phase of understanding it semantically. The process of planning the experiment themselves likely increased the students' commitment to the activity and led them to pay more attention to the details. Consequently, in this lesson, the act of students' describing the experimental plan themselves likely served as a catalyst for considering the purpose and method of the experiment on a semantic level. The results also indicated that many students derived a sense of accomplishment from such experiences and recognized the significance of devising the experimental plan themselves (Oshima, 2020). However, making students devise their experimental plan should be assumed to necessarily lead to the aforementioned outcomes. For instance, if the scientific content is too complex or if the students cannot figure out the factors to determine in the experimental method, it is unlikely they would demonstrate increased commitment to the experimental activity. Oshima (2020) found that, even with a small experimental method component, the responsibility was handed over to the students to decide on the basis of their own thinking. This responsibility created a situation where the students had to reach a semantic understanding phase because the experimental process would not proceed unless they devised the experimental method themselves. This experience led students to make decisions on the basis of their own thoughts and feelings, which is believed to have elicited a positive response.

In the recipe approach, students only need to understand and follow the procedures of the "correct" experimental methods provided by their teachers. By contrast, in this class, students had no correct experimental method to refer to. Instead, they had to evaluate the experimental method themselves and make decisions on the basis of their sense of its validity. This process, described by Germann et al. (1996) as creative, might be done unconsciously or automatically by experts and teachers. However, it is a significant task for students who are still learning (Oshima, 2020). Incorporating elements of decision-making into regular science classes can not only promote decision-making skills but also emphasize the need to consider the content and method from the perspectives of validity and reliability, enabling deeper learning.

Emphasizing the human elements of science enterprise

By focusing on the process of trial and error, including failures and creativity leading to scientific achievement, science education can become a more human activity. This shift away from superficial memorization-based teaching can enable students to develop skills applicable in complex real-life situations.

This section summarizes Huang and Oshima (2021), who developed two lessons that deliver content knowledge while emphasizing elements of the Nature of Science (NOS) as science enterprise. These lessons are designed to simulate the activities performed by scientists in the process of producing scientific knowledge. As Klopper and Cooley (1963) pointed out, historical events in science include realistic understandings about science and scientists and they convey important ideas about science—elements of NOS. Students were expected to conduct practical activities based on these historical events and appreciate the elements of NOS related to the science content. The selected topics for Grade 8 students ($n = 72$; separated in two class groups) were "Chemical reactions" and "Atoms and molecules."

The first lesson unit involved conducting an experiment to verify the erroneous phlogiston theory and how it was superseded by the discovery of oxygen gas. This unit ended with the explicit teaching of the NOS element: scientific knowledge is simultaneously reliable and yet tentative. The second lesson unit involved a brief hands-on activity in which each group of students arranged a set of polygon shapes in any form of their choice. This activity underscored the point that creativity is critical in the production of scientific knowledge as the NOS element.

Results from the pre-test indicated that approximately 80% agreed that scientific knowledge is subject to change although a majority could not provide examples (Huang & Oshima, 2021). Moreover, approximately 90% agreed that creativity (imagination) plays a role in scientific investigations, with approximately 60% of these students specifying that creativity is essential in the planning of an experiment. However, the other aspects of scientific investigations were not as prominently highlighted.

The post-test results showed marginal improvements, with no significant increase in the number of students providing examples of knowledge variability or indicating other aspects of scientific investigations beyond planning. Despite this result, the majority of students found the lesson units to be engaging and interesting because they could understand the concepts through hands-on experiments (Huang & Oshima, 2021).

These lessons exemplify that integrating NOS elements into the daily teaching and learning of science through activities and/or experiments is feasible. Notably, however, a few lessons might not be sufficient for a comprehensive enhancement of student's understanding of NOS. Nevertheless, science lessons can be rendered more meaningful and impactful when they encompass the authentic experiences of scientists, including their trial and error.

Challenges in science education in the STEAM education movement

As shown thus far, the focus of STEAM education in Japan lies in period for inquiry-based cross-disciplinary study and inquiry-based study of science and mathematics in upper secondary school. In these classes, among diverse objectives, students are expected to conduct exploratory initiatives that lack a single solution or answer. However, the current science lessons and the STEAM Library, which is a platform for STEAM education, are considered insufficient in promoting inquiry activities. We have consequently proposed two types of lessons according to Oshima (2020) and Huang and Oshima (2021). The perspective in developing these lessons was the incorporation of human elements such as decision-making and trial-and-error in experimental activities. By introducing the element of "humanity," which is influenced by diverse ways of thinking and experiences, science can be considered an integrated activity in and of itself. We can advocate for rich learning activities where students can become, or indeed must become, the main decision-makers regarding diverse elements, rather than merely following "absolute science." By accumulating such learning experiences, students can effectively engage in full-fledged STEAM learning in upper secondary school.

In such inquiry activities that incorporate "human" and comprehensive elements, students may think in incorrect ways or arrive at wrong conclusions. In traditional learning within existing subjects, too much effort might have been devoted to devising ways to avoid mistakes out of fear of students making them. The teacher's comment in Oshima (2020), "most of the students would probably be unable to describe their experimental plans at all," might reflect this attitude. Students cannot naturally engage in experimental activities on new learning content in a "faultless" manner; they have therefore not been allowed to make judgments or decisions about experimental plans. If "faultlessness" is overly demanded, students will disengage, resulting in a situation where they "can't." This behavior is merely because they "can't" do things "without mistakes," but teachers remove elements where students might make mistakes in order to change the students' "can't" into "can." Although this approach is rational from the perspective of caring for students and making learning activities a scientifically valid process, it is not necessarily rational from the perspective of confronting students with their own thoughts and letting them engage in inquiry as a personal problem-solving exercise.

In the first lesson proposed in Oshima (2020), students were allowed to plan experiments, incorporating the possibility of ending up with less valid experimental activities. As a result, even when allowed to make even the smallest decisions about experimental methods, students were able to face the experimental activities, understand the purpose of the experiment, and actively engage. If only scientifically valid approaches are provided by the teacher, no room exists for students' questions and students cannot even grasp the starting point for inquiry. The present moment, when STEAM education is demanded, is an opportunity to reflect such education. By making student learning truly exploratory in each subject, sound academic abilities that can be applied at the real-world level will be nurtured. The enrichment of learning through the introduction of human elements is possible not only in science but also in the inquiry of any existing subject.

Acknowledgments

This work was supported by the Japan Society for the Promotion of Science KAKENHI Grant Numbers JP15K17395, JP18K13148.

References

Bennett, J. (2003). *Teaching and learning science: A guide to recent research and its applications.* Continuum.

Cabinet Office. (n.d.). *Government of Japan. Society 5.0.* https://www8.cao.go.jp/cstp/english/society5_0/index.html

Germann, P., Aram, R., & Burke, G. (1996). Identifying patterns and relationships among the responses of seventh-grade students to the science process skill of designing experiments. *Journal of Research in Science Teaching, 33*(1), 79–99. https://doi.org/10.1002/(SICI)1098-2736(199601)33:1<79::AID-TEA5>3.0.CO;2-M

Huang, Y., & Oshima, R. (2021). Lessons for content delivery with emphasis on elements of nature of science in the teaching of middle school science. *Bulletin of the Faculty of Education, Chiba University, 69,* 77–82. https://doi.org/10.20776/s13482084-69-p77

Klopper, L. E., & Cooley, W. W. (1963). The history of science cases for high schools in the development of student understanding of science and scientists: A report on the HOSC instruction project. *Journal of Research in Science Teaching, 1,* 33–47.

Lacorte, R. B., Oshima, R., & Iwasaki, H. (2022). Japan's lower secondary school science practices and students' perceptions: A Filipino science teacher's point of view. *Journal of Research in Science Education, 63*(1), 15–31. https://doi.org/10.11639/sjst.A21005

Masumoto, M. (2022). *Keizaisangyousyou no "STEAM Library" ni okeru gakusyu katsudou ni kansuru kenkyu (Research on learning activities in the "STEAM Library" of the Ministry of Economy, Trade and Industry).* Undergraduate Thesis, Faculty of Education, Chiba University [in Japanese].

Ministry of Economy, Trade and Industry (METI). (2020). "Mirai no kyoushitsu" to EdTech kenkyukai STEAM kentou working group chukan houkoku (Interim report of the "Future Classrooms" and EdTech Research Group's STEAM Discussion Working Group) [in Japanese]. https://www.learning-innovation.go.jp/existing/doc202008/steam2020-midreport.pdf

Ministry of Education, Culture, Sports, Science and Technology (MEXT). (2018). *Society 5.0 ni muketa jinzaiikusei~Shakai ga kawaru, manabi ga kawaru~ (Human resource development for Society 5.0: Changes to society, changes to learning)* [in Japanese]. Retrieved from https://www.mext.go.jp/component/a_menu/other/detail/__icsFiles/afieldfile/2018/06/06/1405844_002.pdf

Ministry of Education, Culture, Sports, Science and Technology (MEXT). (2019). *Koutou gakkou gakusyu shidou youryou (Heisei 30 nen kokuji) kaisetsu risu hen (Guidelines for the course of study as national curriculum standard for upper secondary school in period for inquiry-based study of science and mathematics).* Tokyo Shoseki Co., Ltd. [in Japanese].

Ministry of Education, Culture, Sports, Science and Technology (MEXT). (2023). *Koutou gakkou gakusyu shidou youryou (Heisei 30 nen kokuji) kaisetsu sougoutekina tankyu no jikan hen (Guidelines for the course of study as national curriculum standard for upper secondary school in period for inquiry-based cross-disciplinary study).* Gakkotosho Co., Ltd. [in Japanese].

Ministry of Education, Culture, Sports, Science and Technology (MEXT) (Central Council for Education). (2021). *"Reiwa no nihongata-gakkōkyōiku" no kouchiku wo mezashite: Subete no kodomotachi no kanousei wo hikidasu, kobetusaitekinamnabi to kyōdoutekina manabi no jitugen (tōshin) (Toward the construction of "Japanese style school education": Optimal individualized learning and collaborative learning that bring out the potential all children (report)* [in Japanese]. https://www.mext.go.jp/content/20210126-mxt_syoto02-000012321_2-4.pdf

Oshima, R. (2020). Jikken houhou no datousei ni kakawaru handan wo tomonau jikken keikaku ni taisuru seito no ishiki – chugakkou daiichi gakunen "chikara no okisa to bane no nobi" wo jirei toshite – (Students' awareness of experimental planning activities emphasizing judgment of validity of experimental methods: A case study of 'Hook's Law' for grade 7 students). *Journal of Research in Science Education, 61*(2), 219–228 [in Japanese]. https://doi.org/10.11639/sjst.20002

Oshima, R., Fujita, T., Awaya, T., & Pambit, R. (2017). Science teacher's recognition of inquiry-based learning in Japan and the Philippines. *Journal of Science and Mathematics Education in Southeast Asia, 40*(2), 142–158.

Parkinson, J. (2004). *Improving secondary science teaching*. Routledge.

Sageme, M., & Oshima, R. (2023). Relationship between science career awareness and individual interest in school science in Japanese Junior High School Students. *Bulletin of the Faculty of Education, Chiba University, 71*, 269–281. https://doi.org/10.20776/S13482084-71-P269

Tsuruoka, Y. (2017). Rika kyouiku to kyaria kyouiku (Science education and career education). In T. Hashimoto, Y. Tsuruoka, & S. Kawakami (Eds.), *Gendai rika kyouiku kaikaku no tokusyoku to sono gugenka* (pp. 122–130). Toyokan Publishing Company Co., Ltd. [in Japanese].

10 Investigating inquiry-based activities to acquire learning content and competence in STEM education

A case study in chemistry

Kiichi Amimoto

Introduction

Features of the discipline of chemistry and its learning

Chemistry deals with the composition, structure, and properties of matter; the processes that matter undergoes; and the energy changes that accompany these processes (Sarquis & Sarquis, 2012). In the field of natural science, chemistry systematizes the principles of matter in its properties, changes, and behavior through experiments and by associating the structure of matter with the features of its changes. Another important feature of chemistry is that novel materials are produced, and chemical phenomena have practical applications in science and technology based on a fundamental understanding of substances and their changes.

Chemistry is a material-based discipline that contrasts with other disciplines of natural science. Discussing scientific principles for physical phenomena in the natural world has a perspective similar to that of the academic approach to physics. Classifying numerous elements regarding their periodicity and describing the relationships of change among substances based on their structure are similar to biology, which discusses the diversity of living organisms in taxonomic terms. Furthermore, chemistry has significant connections with biology and earth sciences regarding the study of all matter in nature and the universe. With such connections to various disciplines, chemistry is sometimes called the "Central Science," referring to the title of the book *Chemistry: The Central Science* published by Brown and LeMay (1977) which is now up to its 15th edition in 2021. Figure 10.1 presents a schematic diagram showing the relationship between chemistry and other science and STEM disciplines.

Based on these disciplinary features, a learning outcome in chemistry through elementary and secondary science education is required for students to acquire a method of natural recognition that approaches substances by relating their properties, changes, and structures to each other. The knowledge and skills that students acquire in the study of chemistry are necessary for a safe and cultured life and constitute an important part of the scientific literacy

DOI: 10.4324/9781003392545-11

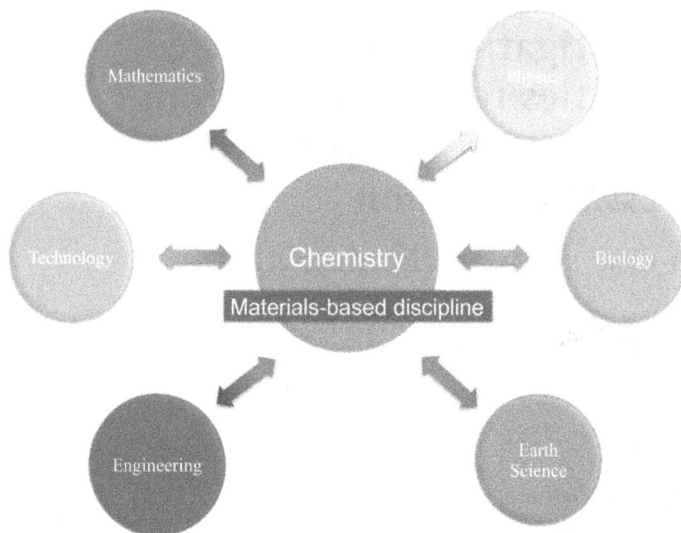

Figure 10.1 Chemistry as central science.

required of a society that aims for sustainable development. In addition, logical thinking developed through chemistry contributes to the deepening of scientific thinking, which enables students to correctly judge and solve various problems in daily life and society.

Developing teaching materials based on the specific features of science learning in Japan

In Japan, the Course of Study for School Education revised in 2018 presents three pillars as the attributes and abilities to be cultivated: "knowledge and skills," "ability to think, judge, and express oneself," and "ability to learn and humanity" (Ministry of Education, Culture, Sports, Science and Technology [MEXT] (Central Council for Education), 2018). The curriculum is organized to improve the learning process qualitatively through "independent, interactive, and deep learning" in all subjects. It is important for students to deepen their scientific understanding and inquiry into natural objects and phenomena. Furthermore, students must realize the usefulness of learning science through observation and experimentation with perspectives and while applying "the way of viewing and thinking in science."

Based on the above background, three important considerations for developing teaching materials in chemistry are as follows:

1 Use appropriate teaching materials and tools to help students acquire solid knowledge and skills regarding chemical concepts. This is related to the development of "knowledge and skills."

2 Students should experience and understand the process of scientific inquiry into natural objects and phenomena through learning activities including the "process of inquiry." These activities consist of recognizing a problem, setting a problem and hypothesis, designing and conducting an experiment, and discussing the results. This is related to the development of the "ability to think, judge, and express oneself."

3 Learning activities in which students understand what they have learned in relation to human life and society will enable them to transform the attributes and abilities acquired through these learning activities into versatile ones. This corresponds to the cultivation of "ability to learn and humanity."

Designing inquiry activities in STEM education

Triad of inquiry activities including content, competencies, and learning contexts

The constructive principle for developing the inquiry activities described above is to program both content- and competence-based learning as threads of context-based learning, in which students understand science in relation to real-life situations. Content-based learning ensures that students learn science content, and competence-based learning cultivates the abilities and attitudes for scientific investigation.

A content-based learning approach is highly effective in ensuring that students systematically learn scientific content. In chemistry, students are expected to learn the structure, change, and properties of matter by finding models and laws from chemical events, understanding the things around them, and applying the acquired knowledge and skills to explore the next event. This leads to a spiral and higher-order learning activity. However, competence-based learning is also desired to cultivate the ability and attitude of scientific exploration through science study and inquiry. In this process, one constructs learning development with perspectives such as "knowing how to learn," "tackling issues independently," "understanding and exploring events based on evidence," and "refining sensitivity through experience." Isozaki (2014) has noted:

In rethinking Japan's science curriculum, redefining learning (competence) from a contextual and social participatory perspective, as well as adding not only scientific knowledge (content) but also knowledge about science, such as how science works, will help students make decisions based on scientific evidence and develop values related to science (competence). To appropriately develop both content and competence, it is also useful to engage with knowledge about science, including how science works in real-life contexts.

Cross-disciplinary inquiry activities with STEM concepts and their affinity with chemistry

STEM education intends to enable students to acquire cross-disciplinary skills by linking related disciplines. Vasquez et al. (2013) and Matsubara and Kosaka

(2017) have described three approaches to cross-curricular learning: thematic, interdisciplinary, and transdisciplinary.

1 In the thematic approach, concepts and skills specific to each subject are taught individually.
2 The interdisciplinary approach teaches deeply related concepts and skills from two or more subjects.
3 Transdisciplinary approaches involve learning to apply knowledge and skills from two or more subject areas by working on real-world tasks and projects.

Understanding these approaches will guide us in designing a step-by-step approach to cross-disciplinary learning that is appropriate for students' actual situations. First, through a thematic or interdisciplinary approach, students create a foundation for developing the core concepts of each discipline and the skills needed to cross them. Based on this learning, a higher-level transdisciplinary approach should be adopted to guide students to the next stage, where they aim to develop the ability to live and work in the real world.

Chemistry is related to other disciplines in the natural sciences and serves as a foundation for science and technology. Therefore, chemistry can provide materials for cross-curricular inquiry activities in other areas of natural science, such as physics, biology, and earth science, as well as in science, technology, engineering, and mathematics. This means that chemistry has the potential to compose inquiry activities centered on it, together with other disciplines. Figure 10.2 illustrates a conceptual diagram that represents the design of inquiry activities in STEM education that allow students to learn content and competence through a context.

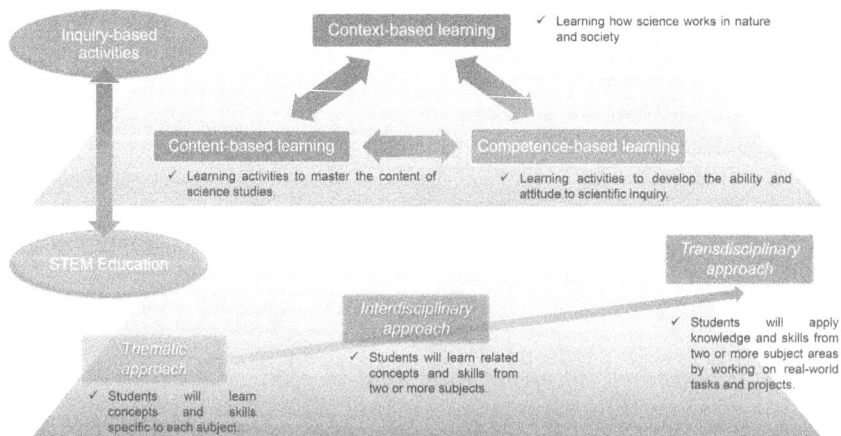

Figure 10.2 Conceptual drawing on the design of inquiry activities.

Examples of inquiry activities in chemistry

Transversal inquiry activities across several units of study in chemistry

In the unit of "Organic Chemistry" in high school chemistry, students learn to understand the structure, properties, and reactions of organic compounds, acquire skills in observation and experiments, and express the results of their exploration. As a specific teaching material that meets such unit objectives, Amimoto and Endo (2014) have introduced the "Identification Experiment of Organic Compounds" using hydroxy acids. Hydroxy acids are organic compounds that are deeply related to our daily lives as sour components of fruits and food additives and can be used as materials for comprehensive experimental activities that introduce the chemical contents of alcohols and carboxylic acids. When four hydroxy acids (lactic acid, malic acid, tartaric acid, and citric acid) are used for identification, only lactic acid shows an iodoform test; oxalic acid produced by the oxidation of tartaric acid is further decomposed to carbon dioxide, and gas generation is observed; and citric acid, which is a tertiary alcohol, is not easily oxidized. The four hydroxy acids can be easily identified from the results of three experiments.

The learning sequence using this material is as follows: (1) introduction, students review the structure and reactivity of hydroxy acids; (2) setting a hypothesis, students predict the results of the iodoform reaction and oxidation reaction with hydroxy acids, and plan a method of identification based on the differences in reactivity (see Figure 10.3); (3) development, students conduct experiments according to the methods planned by the students themselves; and (4) conclusion, students present the identified results along with their thought process.

There are several ways to combine the three experiments to establish a hypothesis. By allowing students to discuss and decide on identification

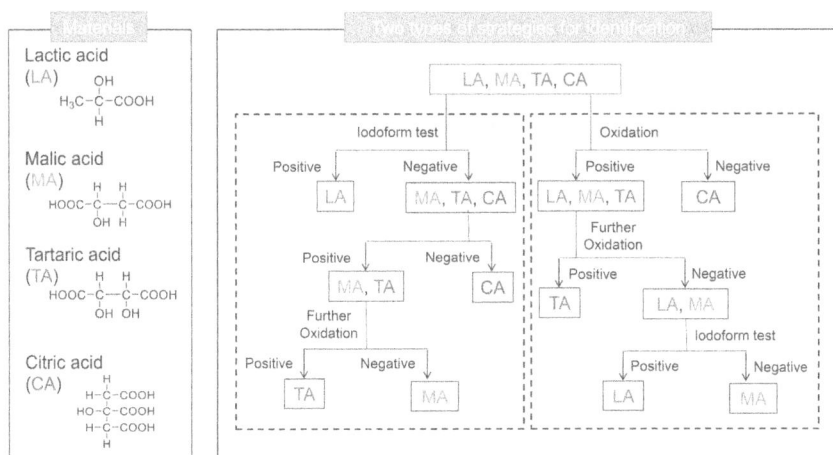

Figure 10.3 Flowcharts of students' strategies for identification of hydroxy acids.

strategies, an inquiry aspect can be incorporated into conventional organic chemistry experiments. Furthermore, although this activity was an experiment in an organic chemistry unit, it effectively allowed students to reflect on their previously learned knowledge and skills, such as oxidation and reduction, and the collection and identification of evolved gases. In this sense, it is interesting that this activity is an example of a transversal inquiry across several units of chemistry.

Studies on its effects in educational practice and teacher training have revealed that many students can derive correct answers and logically explain the reasons for their judgments. The following four points are illustrated: (1) students can engage in inquiry activities with a clear objective of identifying organic compounds; (2) students need to apply scientific thinking by employing their knowledge and skills in organic chemistry and related topics when designing analytical methods to achieve the objective of identification; (3) students can logically interpret the process of identification to draw conclusions; and (4) the achievement that students have been able to derive correct answers in their own way promotes a sense of self-reliance and interest in chemistry. These results show that laboratory activities are effective in further deepening students' understanding of general chemistry content as well as improving their competence in designing experiments and developing an attitude of self-directed problem-solving.

Interdisciplinary inquiry activities using colored materials

STEM-oriented science education requires a learning design that integrates science with technology, engineering, mathematics, and other fields based on learning. Therefore, colored materials are promising chemical substances. Colored materials have been used as pigments and dyes to enrich human life by adding color to our daily lives. The Science field is concerned with the synthesis of pigments and dyes based on their structures and properties, whereas the Technology and Engineering fields are concerned with the manufacturing of colored materials by compositing various substances. While a sense of Art and design is essential to depict the world by color materials, technical methods are used to preserve these works, and the Mathematical treatment of light absorption and reflection is necessary to correctly analyze the degree of coloration. When we look at colorful chemicals in this way, pigments and dyes are materials that are well connected to STEM (or STEAM) disciplines and have the potential to be one of the materials of inquiry in STEM education. In addition, as mentioned earlier in this chapter, an effective method of cross-disciplinary learning in STEM education is to use context-based learning that links mutual connections between disciplines in relation to human life. *Salters Advanced Chemistry* (University of York Project Team, 2015), which consists of a context-based learning curriculum, has a chapter titled "Color by Design" which describes the chemistry of pigments that color our lives and arts, touching on their structure, properties, chemical analysis methods, and dyeing.

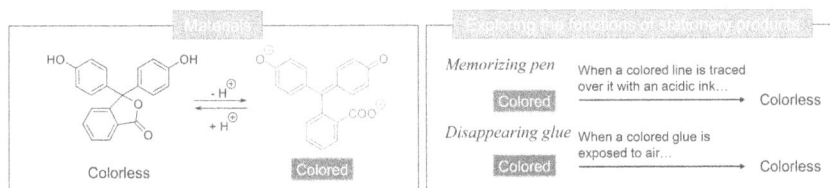

Figure 10.4 Stationery products using pH-dependent color change of phenolphthalein.

An example of the application of dyes in daily life is stationary products. A memorizing pen can be used to erase color when a red marker line is traced over it with another pen. Some glues are colored when applied to paper but fade to transparent when exposed to air. These stationery products effectively utilize the color changes of acid-base indicators (see Figure 10.4). Phenolphthalein, which can be easily prepared by heating phenols and phthalic anhydride under acid catalyst, can be used as a model material to develop exploratory activities on the theme of exploring the chemical functions of these stationery products (Amimoto & Koga, 2009).

The learning process using this material is as follows: (1) introduction, after identifying the pH-dependent color change of phenolphthalein, a teacher mentions that a memorizing pen and a disappearing glue are stationery items that utilize the pH-dependent color change of phenolphthalein; (2) setting a hypothesis, students discuss the working principle of these stationery items based on their knowledge of pH-dependent color changes in acid-base indicators; (3) development, through an experiment in which the red color of a model material prepared by dispersing phenolphthalein in a polymer matrix disappears when an acidic aqueous solution is added, students evaluate their hypotheses regarding the working principle of a memorizing pen; and (4) discussion, analogous to the mechanism of a memorizing pen, students can see that the blue color of the glue disappears when the weakly basic glue becomes acidic under the influence of carbon dioxide in the air.

This activity is designed in the context of the use of chemical substances in stationery items through simple experiments to determine their functional mechanisms. The fact that the acid-base indicator is applied to familiar stationery items, which is unexpected for students, stimulates their interest while also allowing the activity to provide an example of an interdisciplinary inquiry activity regarding the connection between chemistry, science, and technology.

Transdisciplinary inquiry activities to elucidate biological phenomena through the modeling of chemical reactions

Finally, an example of a transdisciplinary inquiry activity examining the relationship between chemistry and food science is presented. Various bio-related substances play a role in the organization of the body and maintenance of homeostasis. Among these, the heme-iron pigment protein is an important biological substance responsible for oxygen transport at the cellular level. Bylkas and Andersson (1997) have reported the purification and spectroscopic analysis of myoglobin extracted from meat. Bailey (2011) has reported an undergraduate experiment regarding ligand-binding reactions of myoglobin with carbon monoxide (CO) and oxygen (O_2). Motivated by these previous studies, Amimoto and Ise (2013) have reported on inquiry activities that allow students to investigate the function of myoglobin and its relationship with life and food.

Myoglobin is readily extracted from beef rumps in a phosphate buffer solution. Using this solution, experiments were conducted to visualize the O_2-binding ability of myoglobin. The experiment procedure is as follows: (1) the solution immediately after extraction is bright red, indicating that the heme iron ion present is Fe(II), which forms an oxygen complex upon contact with O_2; (2) when a small amount of an oxidizing agent such as hexacyanide iron(III) acid ion $[Fe(CN)_6]^{3-}$ is added to this solution, the heme iron ion is oxidized to Fe(III) and becomes unable to combine with O_2, and the color of the solution changes from red to yellow; (3) when a reducing agent such as sodium hydrosulfite $Na_2S_2O_4$ is added to the solution, the heme iron ions are reduced to Fe(II), and the color of the solution changes from yellow to reddish-purple; finally, (4) when O_2 is bubbled into the solution, O_2 is co-ordinated to the heme iron ion, and the color of the solution changes from reddish-purple to bright red. Thus, the solution returns to its initial state, as described in (1). Figure 10.5 shows the processes involved in the transport of oxygen by heme iron.

Heme iron bound to O_2 transports oxygen and dissociates it as required. Heme iron, having lost O_2, recombines with O_2 under O_2 saturation. Arterial blood, which transports oxygen from the heart to each organism, is bright red, whereas venous blood, which returns to the heart from each organism, is

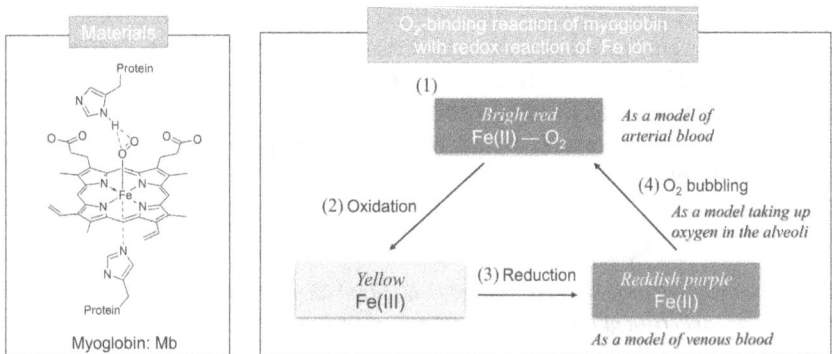

Figure 10.5 Chemical equilibria of O_2-binding reaction of heme iron in steps (1) to (4).

dull red. Experiments (1)–(4) correspond to the model experiments in which the function of heme iron and the color change of blood are visualized in a test tube. In the model experiment described above, arterial blood corresponds to the solution in (1), and venous blood corresponds to the solution in (3) with heme-Fe(II) under anaerobic conditions. Furthermore, (4) suggests that venous blood takes up oxygen from the alveoli to become arterial blood. A fascinating aspect of this teaching material is that students can scientifically discuss the function of heme iron as an O_2 carrier by considering the commonly known color of blood in connection with the valence and coordination structure of heme iron.

This study can be extended to further developmental activities that allow students to explore the functions of food additives in meat preservation. When sodium nitrite $NaNO_2$ is added to the solution in (3) of the above experiment, a stable nitroso complex is formed, where the nitroso ion NO^- is coordinated to the heme iron ion, changing the color of the solution to pink. Bubbling O_2 into the solution hardly changes the solution's color. This indicates that heme iron no longer functions as an O_2 carrier because NO^- cannot be replaced by O_2 (see Figure 10.6). An important implication of these findings is the function of $NaNO_2$ as a food additive, both as a preservative and coloring agent. In other words, meat treated with $NaNO_2$ shows increased food preservation because heme iron is unable to transport O_2, thus preventing the survival of O_2-requiring bacteria. After this treatment, the meat becomes pink, indicating that it also serves as a chromogenic agent.

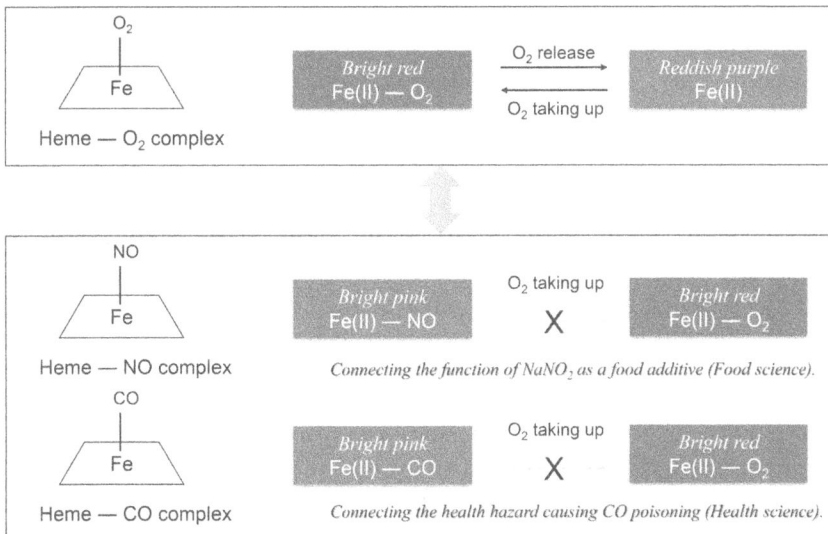

Figure 10.6 Heme iron bound to nitroso ion or carbon monoxide is no longer exchanged for oxygen.

The learning process using this teaching material is as follows: (1) introduction, using a solution of myoglobin extracted from meat, students observe the redox and O_2-binding ability of heme iron as a color change in the solution; (2) development, from the spectral analysis, students can understand that when a myoglobin solution is treated with $NaNO_2$, it loses its O_2-binding ability. Based on this result, the students discuss the function of $NaNO_2$ as a food additive; and (3) expression, activities are designed to scientifically explain the topics of food preservation and health science that students already know about in their daily lives by correlating their understanding of the O_2-binding ability of heme iron. An example is asking the following questions:

Q: When carbon monoxide (CO) produced by the incomplete combustion of organic substances is inhaled, people suffer from CO poisoning, which sometimes leads to death. Treatment for CO poisoning sometimes involves the administration of 100% O_2 through an oxygen mask or hyperbaric oxygen therapy. Patients suffering from CO poisoning appear to have good blood complexity. Explain these causes or reasons related to CO poisoning.

A: Assuming that heme iron bound to CO with the isoelectronic structure of NO– becomes incapable of carrying O_2, students explain that the O_2 partial pressure must be increased to shift the chemical equilibrium of ligand exchange from CO to O_2 on the heme-iron complex, and the light red color of the CO-heme iron complex makes the patients' blood look better.

This series of inquiry activities is an example of a transdisciplinary activity that links chemistry and biology in science, as well as technology and engineering through food processing, health science, and medical science. From the standpoint of developing competence, it is also interesting to highlight Nature of Science (NOS), which refers to how scientific principles can be applied to the daily lives of the community.

Conclusion

In this chapter, an overview of the features of the discipline and learning of chemistry and characteristics of science learning in Japan was described. This was followed by examples of triadic inquiry activities involving content, competence, and learning contexts related to real life using the framework of STEM education and focusing on the results of the author's research. In Japan, learning developments with inquiry processes are perceived to be the same as those in STEM education. Isozaki and Isozaki (2021) have described the importance of collaboration among researchers and practitioners from various disciplines in STEM education with an understanding of the purpose and framework of STEM to advance Japanese-style STEM education. Further advances in the development of teaching materials, inquiry-based activities, and their practical application, as a complementary effort to the theoretical research in the previous chapters of this book, will lead to the widespread adoption of Japanese-style STEM education.

Acknowledgment

This work was supported by the Japan Society for the Promotion of Science KAKENHI Grant Numbers JP21H00919 (JP23K20744), JP22K03003.

References

Amimoto, K., & Endo, D. (2014). Kōtōgakkou no chishiki-rikai wo sōgōteki ni katsuyōsuru hidorokisisan no shikibetu-jikken (Identification experiment of hydroxy acids for high school students based on their organic chemistry concepts). *Journal of Science Education in Japan*, *38*(4), 220–227 [in Japanese]. https://www.jstage.jst. go.jp/article/jssej/38/4/38_220/_article

Amimoto, K., & Ise, A. (2013). Miogurobin no kinō to seimei-shokuhin tono kanren wo tankyusaseru kagaku-jikken (A chemical experiment to explore the function of myoglobin and its relationship to life and food). In *Proceedings of CSJ West Japan Chemistry Forum 2013 in Hiroshima*, 196. [in Japanese].

Amimoto, K., & Koga, N. (2009). Phenolphthalein as organic teaching materials: Small-scale preparation and modeling for some functional dyes. *Chemical Education Journal*, *13*(1), Reg. No. 13–11. http://www.edu.utsunomiya-u.ac.jp/chem/v13n1/11_2d4_2.pdf

Bailey, J. A. (2011). An undergraduate laboratory experiment in bioinorganic chemistry: Ligation states of myoglobin. *Journal of Chemical Education*, *88*(7), 995–998. https://pubs.acs.org/doi/10.1021/ed100600c

Brown, T. L., & LeMay, H. E. (1977). *Chemistry: The central science*. Prentice Hall.

Bylkas, S. A., & Andersson, L. A. (1997). Microburger biochemistry: Extraction and spectral characterization of myoglyobin from Hamburger. *Journal of Chemical Education*, *74*(4), 426–430. https://pubs.acs.org/doi/10.1021/ed074p426

Isozaki, T. (2014). Rika-kyōiku niokeru gakuryokukan no saikō – hikakukyōikushi teki apurōchi karano shisa (What is this thing called 'gakuryoku' (academic ability) in 'Rika' (school science) in Japan? Rethinking perspectives on 'gakuryoku' through analysis based on comparative history). *Journal of Research in Science Education*, *55*(1), 13–26 [in Japanese]. https://www.jstage.jst.go.jp/article/sjst/55/1/55_sp13010/_article/-char/en/

Isozaki, T., & Isozaki, T. (2021). Nihon-gata STEM kyōiku no kouchiku ni mukete no rironteki-kenkyū – Hikaku-kyōikugakuteki-shiza kara no bunseki wo tōshite (Theoretical research for establishing a Japanese-style STEM education: Analysis for a comparative historical point of view). *Journal of Science Education in Japan*, *45*(2), 142–154 [in Japanese]. https://www.jstage.jst.go.jp/article/jssej/45/2/45_142/_article/-char/en

Matsubara, K., & Kosaka, M. (2017). Shishitsu-nokyoku no ikusei wo jūshisuru kyōka-ōdanteki na gakushu toshiteno STEM-kyōiku to toi (A discussion of STEM education and questions fostering competencies in the Japanese curriculum). *Journal of Science Education in Japan*, *41*(2), 150–160 [in Japanese]. https://www.jstage.jst. go.jp/article/jssej/41/2/41_150/_article/-char/en/

Ministry of Education, Culture, Sports, Science and Technology (MEXT) (Central Council for Education). (2018). *Koutou gakkou gakusyu shidou youryou (Heisei 30 nen kokuji kaisetsu rikahen resuhen)* (*Teaching guidance of the course of study for Upper Secondary School*) [in Japanese]. Retrieved from https://www.mext.go.jp/content/20211102-mxt_kyoiku02-100002620_06.pdf

Sarquis, M., & Sarquis, J. L. (2012). *Holt McDougal modern chemistry.* Houghton Mifflin Harcourt Publishing Company.

University of York Project Team. (2015). Chapter 10. Colour by design. In *A level salters advanced chemistry for OCR B* (4th ed., pp. 490–551). Oxford University Press.

Vasquez, J., Sneider, C., Comer, M. (2013). *STEM lesson essentials, grades 3–8: Integrating science, technology, engineering, and mathematics.* Heinemann.

11 STEM/STEAM approach in biological education in Japan

Ko Tomikawa

Introduction

Biology is the field of science most closely related to human life, and biological development is driven largely by innovative technologies. For example, Robert Hooke (1635–1703) used a microscope to observe various microscopic organisms and discover cells, which are the basic unit of life. Microscopes have made dramatic progress since the latter half of the nineteenth century and have greatly contributed to developments in biology right up to the present day.

The innovative methods that have dramatically advanced biological research in recent years depend on the development of analytical instruments and other technologies. The development of next-generation sequencing methods has facilitated whole-genome analysis, and biological research has progressed with the development of bioinformatics. Furthermore, metagenomic analysis has made it possible to analyze the genome of an entire environment and obtain detailed genomic information on unknown species. Additionally, the development of genome-editing technology has enabled highly accurate breeding.

Over the past 20 years, the development of innovative methods in biology, mainly in the United States, has fundamentally changed the way humans think about biological research. Modern biology is an extremely complex and multidisciplinary field. This development can be attributed to the fact that biology is closely related to human health and medicine and is the ultimate applied study of human nature.

With such rapid advances in technology, the need to incorporate various methods into the field of biology education has been discussed (Munn et al., 1999), highlighting the usefulness of science, technology, engineering, and mathematics (STEM) approaches (Ari & Meço, 2021). STEM education, along with changes in society, has expanded its concepts since its early days, and the content of educational practices has diversified compared to the early years (Arai, 2018; Hasanah, 2020). Additionally, STEAM education is now provided with the addition of the arts (Liberal Arts) to develop creativity, which is difficult with AI, and to cultivate the ability to verbalize and express one's own images and ideas (Sousa & Pilecki, 2013; Tsujiai & Hasegawa, 2020).

DOI: 10.4324/9781003392545-12

STEM, which focuses on human development in STEM fields, has long been emphasized as a national strategy in the United States and Europe (Isozaki & Isozaki, 2021). Since 2011, the United States has divided the content of science-related studies in school education into four areas: physical science, life science, earth and space sciences as well as engineering, technology, and applications of science. Sciences and engineering are integrated in all educational stages, and cross-curricular learning must be achieved within each science subject (Kosaka & Kumano, 2021; National Science Foundation, 2013). In the United States, molecular biology experiment kits (Bio-Rad "Biotechnology Explorer™ Program") have been developed to enable high school teachers, university faculty, and companies to collaborate in STEM education. Educational material for STEM/STEAM education that emphasizes scientific thinking is becoming increasingly widespread. However, little is known about STEM/STEAM education in biology in Japan, and there little research has been done on the development of teaching material.

Research aims, questions, and methodology

This study's aim is to propose measures to promote modern biology education by incorporating the STEM/STEAM perspective into Japanese biology education. To achieve this objective, the following research questions are presented: (1) What elements of STEM/STEAM education can be found in Japanese courses of study? (2) What kind of practical research on STEM/STEAM education is being conducted in the biological sciences in Japan? (3) How can instructional material for STEM/STEAM education be developed in the biological sciences? The methodology of this study is based on an analysis of administrative documents from the Ministry of Education, Culture, Sports, Science, and Technology (MEXT) and a literature review of previous studies. In addition, new ways to develop educational material are discussed, using our recent laboratory work to develop STEM/STEAM educational material in the biological sciences as an example.

Results and discussion

Elements of STEM/STEAM education found in the courses of study in Japan

Although engineering at higher-level education has historically been well developed in Japan (Isozaki & Isozaki, 2021), science, technology, engineering, and mathematics have been separated, and cross-curricular learning has not been optimally realized. The latest revised courses of study in Japan do not include a description of STEM/STEAM education. Although no STEM/STEAM equivalent exists, the elements of STEM/STEAM education can be found in the courses of study. Therefore, STEM/STEAM education can be implemented by integrating these elements, setting the tasks to be addressed, and designing lessons to address them explicitly (Matsubara & Kosaka, 2017). In this section, we first provide an overview of the relationship between STEM/STEAM education and the curriculum guidelines for science (biology subjects) in Japan as indicated by the MEXT.

In addition to math, elementary school science is related to STEM/STEAM education (Arai, 2019). The concept of STEM/STEAM education can also be found in the science section of elementary school curriculum guidelines (Ministry of Education, Culture, Sports, Science, and Technology [MEXT], 2017a). In addition, the MEXT mentions the need to improve the educational environment, including the enhancement of experimental and observation equipment as well as information and communication technology (ICT) environments (MEXT, 2017a).

Experiments and observations have been emphasized in lower secondary and elementary schools (MEXT, 2017b). The study of biology in lower secondary schools requires students to consider the relationship between biology and human life, particularly the contribution of biology to medicine, industry, and other areas; this is consistent with the purpose of STEM/STEAM education.

Many descriptors are conscious of the qualities and abilities developed through the process of inquiry in the courses of study of biology in upper secondary schools (MEXT, 2018b). In addition, the program incorporates recent rapid advances in life sciences, makes students aware of their relevance to daily life and society, and emphasizes the importance of setting up their own challenges by increasing their interest in living organisms and biological phenomena. This approach is the same as that in STEM/STEAM education, which seeks to solve real problems in society through the process of inquiry and thinking about biology.

Descriptions of STEM/STEAM in biology education in elementary, lower secondary, and upper secondary schools can be found throughout the course of study. As the curriculum of the courses of study for upper secondary school emphasizes the "development of qualities and abilities from a cross-curricular perspective" (MEXT, 2018a), the concept of STEAM education that links biology subjects with other subjects will become even more important in the future.

Practical research on STEM/STEAM education in the biological sciences in Japan

The reports on practical examples of STEM/STEAM education in the biological sciences are few, and even in the United States, only environmental education has been reported (Okumura & Kumano, 2016). This can be attributed to the cutting-edge fields in biology, such as biotechnology, which have made rapid progress in recent years. The content taught in school education up to upper secondary school has become significantly more specialized, requiring teachers to have a precise understanding of the content. The majority of content in lower and upper secondary schools is too advanced to be taught as part of biology classes. In lower and upper secondary schools, the content is too advanced to be handled as part of a biology class, and it is difficult to conduct hands-on experiments (Okumura & Kumano, 2016).

Okumura and Kumano (2016) present two directions for STEM education in biology. The first is STEM education, which does not directly manipulate organisms artificially but rather uses artifacts to create new structures and functions (Direct Modification Artificially; DMA). The second is STEM education, which does not directly manipulate organisms but teaches students to utilize objects while using some kind of artifact (Indirect Modification with Artifact; IMA). To the best of the authors' knowledge, no practical studies have been conducted on STEM/STEAM education using DMA. In contrast, IMA is easier to implement than DMA because it does not directly manipulate living organisms (Okumura & Kumano, 2016). Based on this, Okumura and Kumano (2016) conducted an inquiry activity on the "embryonic development of birds" in an upper secondary school biology class. The results revealed that (1) students expanded their biological knowledge through independent learning activities and (2) problem-solving thinking from a cross-disciplinary perspective of developing a new egg incubation device was triggered.

In STEM/STEAM education practice, project-based learning (PBL) is an effective teaching method (Sawyer, 2014). In the United States, PBL is actively practiced and has proven to be a central means of STEM education (Bradley-Levine & Mosier, 2014). In Japan, PBL is also considered important as an educational method for "active learning" promoted by the MEXT ("independent, interactive, and deep learning" in the courses of study). Okumura and Kumano (2018) further developed the aforementioned practice using "avian embryonic development" as a subject and conducted a study with students of upper secondary school. Okumura and Kumano (2018) conducted a PBL-based practical study of "artificial hatching of baby chicks" for students of upper secondary school, expanding on the previous study on the "development of avian embryos." The results revealed that students could answer the driving question regarding hatching chicken chicks using biological (S in STEM education), engineering (E in STEM education), mathematical (M in STEM education), and technological (T in STEM education) ideas. This suggests that PBL practice may have led to the expansion of cross-disciplinary thinking from the perspective of STEM/STEAM education.

However, in domains such as ecology and evolution, which deal with large-scale temporal and spatial issues, learning based on actual experience is difficult. Shingai et al. (2021) developed a simulation-based digital game for biodiversity conservation and conducted a practical study on STEM education using PBL for university students. The proposed issue is the management of *satoyama*, woodlands in which biodiversity is maintained through human management. The results of the PBL program using a digital game revealed that learners were able to acquire knowledge about *satoyama* and learn different management methods for each *satoyama*. In other words, simulation-based digital games are effective in STEM/STEAM education in fields that address large-scale temporal and spatial issues.

Yamashita and Nomura (2017) developed a science lesson in a lower secondary school science unit on "Life-Sustaining Functions," which utilized a STEM framework to contextualize the development of artificial heart valves. A newly developed model teaching material was proven useful for STEM education in connecting textbook knowledge of blood circulation with the development of artificial heart valves.

In summary, practical research on STEM/STEAM education in the biological sciences in Japan has been conducted by employing various approaches that utilize IMA, simulations, and models and are based on PBL to varying degrees. Matsubara and Kosaka (2017) found that the qualities and abilities to be fostered, the questions to be asked, and the teachers' role differ depending on the degree of integration in STEM. In future, it will be necessary to evaluate the degree of STEM integration in each study.

Future STEM/STEAM education developments in biological sciences – reexamination of existing study contents

As mentioned above, the reported examples of STEM/STEAM educational practices in the field of biology are few. However, as many examples of content currently covered in Japanese biology education include a STEM/STEAM educational perspective, it is possible to re-examine existing learning content and develop it for STEM/STEAM education. In this section, I attempt to organize content mainly in lower and upper secondary schools from the viewpoint of STEM/STEAM education.

In the "Metabolism" unit, students can learn about enzyme reactions, respiration, and photosynthesis as part of STEM/STEAM education. They learn that enzymes act as catalysts to promote chemical reactions and that the reaction rate varies depending on temperature and pH value. Students can design experimental systems to control the temperature conditions (T and E in STEAM) and calculate and graph pH values (M in STEAM) to predict and experimentally confirm the properties of the enzymes involved in the efficient progression of various chemical reactions in living organisms. As enzymes are closely related to human life and are mentioned in the curriculum guidelines for home economics in upper secondary school, cross-curricular studies in home economics are also possible (MEXT, 2018f; Takagi et al., 2001). For respiration, experiments are often conducted to measure the relationship between alcohol fermentation and temperature in relation to enzyme properties. For photosynthesis, the activities that can be considered include designing devices to measure light intensity and the absorption of oxygen and carbon dioxide using various green plants (T and E in STEAM), graphing these relationships, calculating the amount of photosynthesis (M in STEAM), and estimating the optimal environment for each species (Iida & Katayama, 2021).

For osmotic pressure, which is treated in the "Internal Environment" unit, it is possible to create experimental apparatuses and conduct experiments on osmotic adjustment of body fluid using freshwater shrimp *Neocaridina denticulata* (T and E in STEAM; Taniguchi, 2023, https://www.shinko-keirin.co.jp/keirinkan/kou/science/kagaku-jissen/202101/; Tomikawa & Mukuda, 2023). Oxygen dissociation curves and urine production are closely related to mathematics. For example, learners can consider adapting to different environments by creating oxygen dissociation curves and calculating the amount of oxygen in different animal species; similarly, urine production is an exploratory activity involving calculations and graphing. As oxygen uptake into the body is related to respiration, which is covered in "Health and Physical Education" in lower secondary and upper secondary school (MEXT, 2017d, 2018e), it is possible to develop the project into a cross-curricular study with "Health and Physical Education."

Mendel's laws of heredity, which are dealt with in the "Heredity" unit, establish the combinations of genes when they are passed on to children and are closely related to mathematics. Here, an activity utilizing computer simulations with tablets was considered (T, E, and M in STEAM; Sato et al., 2018). The DNA extraction experiment will lead to the practice of STEAM education by developing learning methods, such as the measurement of DNA concentration by measuring absorbance using a UV spectrophotometer (T and E in STEAM) and the estimation of molecular weight by electrophoresis using an agarose gel (T, E, and M in STEAM; Isaji & Matsumoto, 2005). DNA electrophoresis and molecular weight estimation are also employed in DNA cleavage experiments using restriction enzymes in the "Biotechnology" unit (Sonoyama, 2020) and in molecular phylogenetic analyses in the "Evolution" unit. Human genetics is also covered in health and physical education in lower and upper secondary schools (MEXT, 2017d, 2018e); therefore, this study can be expanded to cross-curricular studies.

In the field of evolution, learning using computer simulations of genetic flotation and microevolution (T, E, and M in STEAM; Jones & Laughlin, 2010) can be considered. In addition, based on the concept of the molecular clock, it is possible to estimate phylogenetic relationships and branching ages using the number of differences in amino acid and DNA sequences. In recent years, phylogenetic analysis software has been developed, many of which are free of charge. It is also possible to create phylogenetic trees using analysis software and molecular data from databases (T and M in STEAM; Kazama et al., 2014; Yamanoi et al., 2012).

The marking-and-recapture method, which is used in the field of ecology, estimates the size of a population and population density using statistical methods and is closely related to mathematics. In addition, because markers must not be easily lost in the field, it is necessary to consider the selection and use of markers with suitable components depending on the habitat of the target organisms (T and E in STEAM). The study of production structure diagrams, which is used to understand the relationship between the structure of plant communities and material production, also includes the measurement of illuminance in plant communities (T and E in STEAM) and the calculation of the relative illuminance between the inside and outside of the leaf layer (M in STEAM), which can be implemented (T and E in STEAM) and used in STEM/STEAM education (Kitagaki & Andoh, 2007). As the production structure diagram is closely related to agriculture, this study can be linked with upper secondary school civics by relating it to securing food security. The study of bioaccumulation requires calculating the concentrations of specific substances at each step of the food chain (M in STEAM). Bioconcentration is also closely related to pollution issues addressed in lower and upper secondary school social studies (MEXT, 2017c, 2018c), and cross-curricular learning with social studies is possible (A in STEAM). As the social studies curriculum guidelines for lower secondary and upper secondary school also mention biodiversity and the Convention on Biological Diversity (MEXT, 2017c, 2018c), it is possible to link the study of this content with social studies. Table 11.1 presents key learnings that could be developed as part of STEM education in biology.

Attempt to develop STEM/STEAM educational material in the field of biology

Recent technological developments have provided various research methods in biology. Biology education deals with the research results obtained by modern research methods.

In recent years, biotechnology and geographic information systems (GIS) have made remarkable progress and are expected to be used in biology education. Here, we introduce examples from our recent work on the development of STEAM teaching material using biotechnology and GIS.

DNA barcoding and molecular phylogenetic analysis of mammals with meat-using biotechnology

Since the dawn of time, humans have used biotechnology to create fermented and preserved foods and to breed useful plants and animals. With the advances in life science research, the structures of genes and DNA have been elucidated. Furthermore, the development of technologies such as cell fusion and genetic recombination has led to the use of biotechnology in various fields, including medicine and agriculture. Biotechnology is expected to make a substantial contribution to human life as a useful technology for solving global issues, such as environmental problems. Biotechnology is closely related to biology and has a long history, thus providing excellent material for STEAM education in the field of biology.

In recent years, the mislabeling of food products has become a social problem. For example, in Japan, a major problem arose in 2007 when a food processor was found to mix ground meat labeled as 100% beef with pork and chicken. In Europe, a problem arose in 2013 when a food product claimed to be beef was mixed with horse meat. Mammals, on the other hand, are an important taxon in terms of biological evolution and are frequently used as material for studying evolution from lower secondary to upper secondary school.

Table 11.1 Key learnings that could be developed as STEM education in biology. Circle indicates the appropriate STEM item for each study content.

Frame	Unit	Learnings	Science	Technology	Engineering	Math	Subjects related to biology
Structure and function of living organisms	Metabolism	Characteristics of enzyme reaction	○	○	○	○	Home economics
		Respiration	○○	○○	○○	○○	
		Photosynthesis	○○	○○	○○	○○	
	Internal environment	Regulation of osmotic pressure of bodily fluids	○				
		Oxygen dissociation curve	○			○	Health and physical education
		Production of urine in the kidneys	○			○	
Continuity of life	Heredity	Mendel's laws of heredity	○	○	○	○	Health and physical education
		DNA	○	○	○	○	Health and physical education
		Biotechnology	○	○	○	○	Civics, Health and physical education, Social studies
	Evolution and phylogeny	Microevolution	○○	○○	○○	○○	
		Genetic drift	○○	○○	○○	○○	
		Molecular phylogenetic analyses	○	○	○	○	
Relationships between living organisms and environments	Ecology	Marking-and-recapture method	○	○	○	○	
		Production structure diagram	○	○	○	○	Civics
		Biological concentration	○			○	Social studies

Species (meats)
↓
DNA extraction
↓
PCR
↓
Electrophoresis
↓
DNA sequencing
↓
DNA barcoding
↓
Constructing
phylogenetic trees

Science
• Food and Agricultural Science
• Mammology
• Molecular Biology
• Phylogeny

Technology
• Biotechnology
• Instrument-based (pipetting, polymerase chain reaction, horizontal gel electophoresis)

Engineering
• Genetic Engineering

Mathematics
• Calculating genetic distances
• Constructing molecular phylogenetic trees
• Determining molecular weight

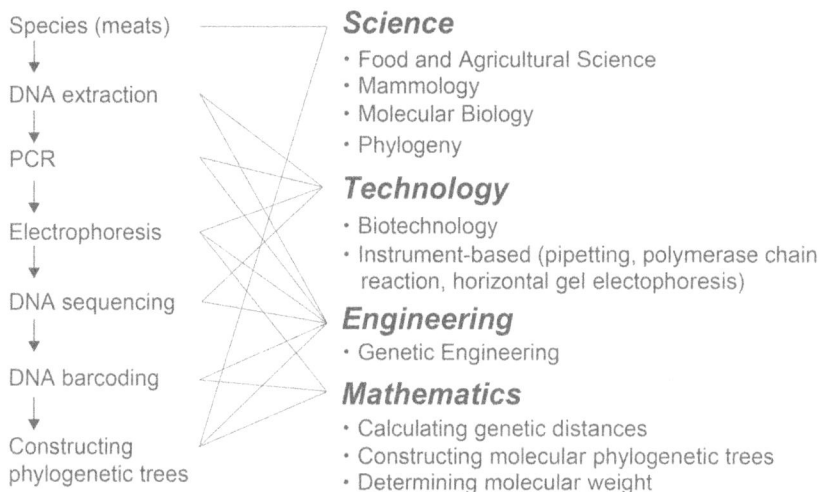

Figure 11.1 The flow of utilization as a teaching aid: DNA barcoding and molecular phylogenetic analyses of mammals using meat.

Thus, we are developing STEAM educational material that uses DNA barcoding to identify meat species in the market and study mammalian evolution based on the obtained DNA sequence data (Minote & Tomikawa, unpublished data). Table 11.2 and Figure 11.1 present the relationship between this educational material and STEAM education.

In developing the teaching material, we first considered the species of mammalian meat that should be used. We focused on the phylogenetic relationships among even-toed ungulates, odd-toed ungulates, and cetaceans and conducted taxon sampling focusing on even-toed ungulates (cattle and pigs), odd-toed ungulates (horses), and cetaceans (whales). The design of suitable primers for polymerase chain reaction (PCR) is essential for DNA sequencing. In this study, we used primers for the mitochondrial cytochrome *b* gene proposed by Sengoku et al. (1993). The results indicated that these primers are applicable to a wide range of vertebrate taxa, ranging from mammals to fish. DNA sequences of rhinoceros and tapirs, which are difficult to sequence, were obtained from databases and added to the analysis. We identified meat species through the homology search program BLAST, using sequences in the database that were highly homologous to the obtained DNA sequences. We confirmed that the meat species used in the analysis could be reliably identified with a BLAST search. Next, we performed molecular phylogenetic analysis to infer the phylogenetic relationships among species. Although many different software packages for phylogenetic analysis were available, we used the freely available MEGA software (Tamura et al., 2021). Although the results of our phylogenetic analyses were largely consistent with previously reported phylogenetic relationships among mammals, a few taxa could not be fully analyzed (Figure 11.2A). This suggests that genetic information in addition to the cytochrome *b* gene is required to accurately estimate phylogeny.

Table 11.2 Proposed learning materials and activities related to STEAM education.

Learning material	Science	Technology	Engineering	Arts	Mathematics	Subjects related to biology
DNA barcoding and molecular phylogenetic analyses of mammals using meat	Food and Agricultural Science, Mammalogy, Molecular Biology, Phylogeny, Taxonomy	Biotechnology, Instrument-based (pipetting, polymerase chain reaction, horizontal gel electrophoresis)	Genetic Engineering	Presentation, preparation of report	Calculating genetic distances, constructing molecular phylogenetic trees, determining molecular weight, graphing	Civics
Environmental education materials integrating habitat analyses and environmental assessment using GIS	Conservation Biology, Ecology, Entomology	GIS technology	Geography and environmental engineering	Presentation, preparation of report	Multivariate analysis, multiple regression analysis	Geography

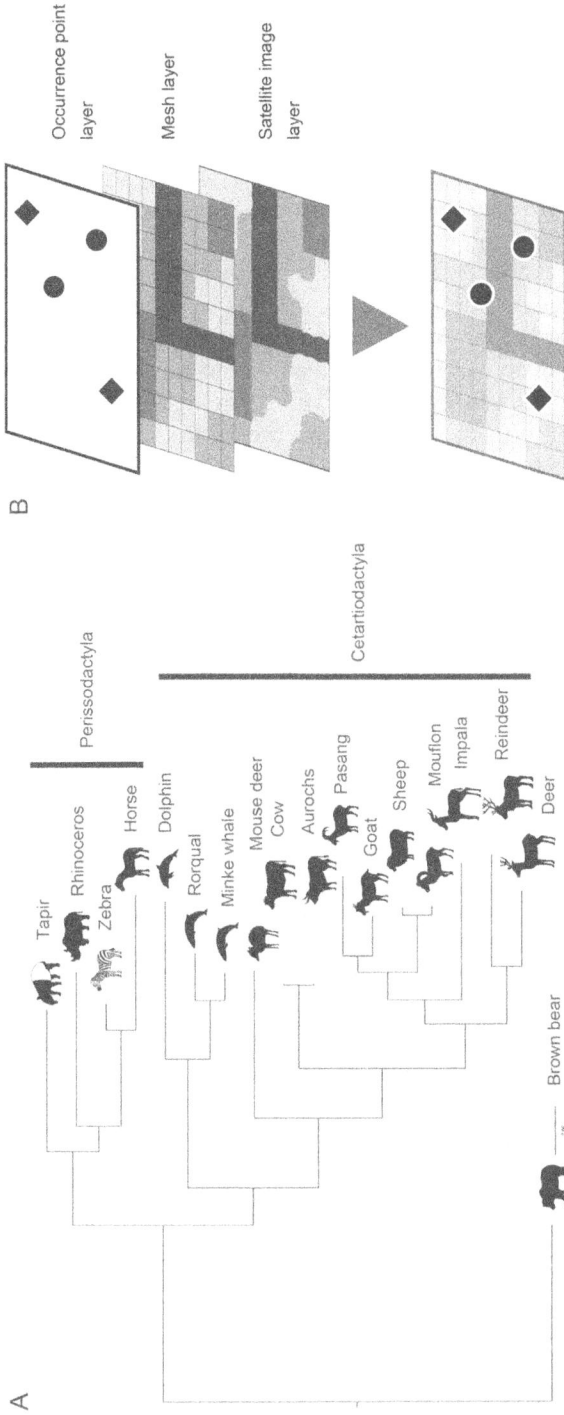

Figure 11.2 (A) An example of molecular phylogenetic tree based on cytochrome *b* gene sequences. (B) Image of the integration of the occurrence point, the mesh, and the satellite image layers in the GIS analysis.

Students learn the skills required to extract DNA from meat, amplify a partial region of the mitochondrial cytochrome *b* gene using PCR, analyze PCR products using electrophoresis, determine the DNA sequence, and compare the sequence with other sequences in the database. This combination of skills enables students to relate scientific questions and techniques in their daily lives. For example, they are expected to make judgments about mislabeling and the unauthorized distribution of meat sold in the marketplace and engage in in-class discussions and presentations.

As food safety and consumer trust are also covered in upper secondary school civics (MEXT, 2018d), cross-curricular learning in the civics field is possible. Furthermore, phylogenetic analyses of the obtained DNA sequence data, together with the sequences of mammalian species in the database, allowed us to infer the phylogenetic relationships among the mammalian taxa (Figure 11.1). It is possible to discuss the evolution of traits in relation to different habitats, such as the evolution of odd-toed ungulates with fewer hooves as they move into grassland environments. In a homology search for DNA sequences, students can learn the statistical background of a homology search.

Furthermore, it is possible to infer the origin of currently domesticated species based on their genetic relatedness. Activities such as hypothesizing about the domestication of mammals, class discussions, and presentations are expected. Because this learning program targets mammals that are used as meat worldwide, it can be applied not only in Japan but also in various countries and regions.

Development of environmental education material using geographic information systems (GIS)

In recent years, biodiversity loss and species extinction due to development and other causes have become global issues. Therefore, the importance of environmental assessment, which is the preliminary investigation, prediction, and evaluation of the environmental impacts of development projects, has been increasing. Against this background, an understanding of the necessity of environmental assessment is required for the study of biology in upper secondary schools (MEXT, 2018b).

A GIS is a technology that visualizes geographic information on digital maps and enables the analysis and display of information relationships, patterns, and trends. GIS has mainly been used in the field of geography but is now used in various fields, such as disaster management and environmental protection, by overlaying various data on a map. GIS has proven to be a powerful tool for wildlife habitat analysis (Ueno & Kurihara, 2014).

We are developing environmental education material that integrates habitat analyses and environmental assessments using GIS, using butterflies that exist on university campuses as examples (Tokuno & Tomikawa, unpublished data). Table 11.2 and Figure 11.3 present the relationship between this educational material and STEAM education.

Field work   ———————→   *Science*

↓   · Conservation
· Ecology
Species identification   · Entomology

↓   *Technology*

· GIS technology
Mapping   · Computer graphics

↓   *Engineering*

· Geography and
GIS analyses   environmental engineering

↓   *Mathematics*

Environmental assessment   · Multivariate analysis
· Multiple regression analysis

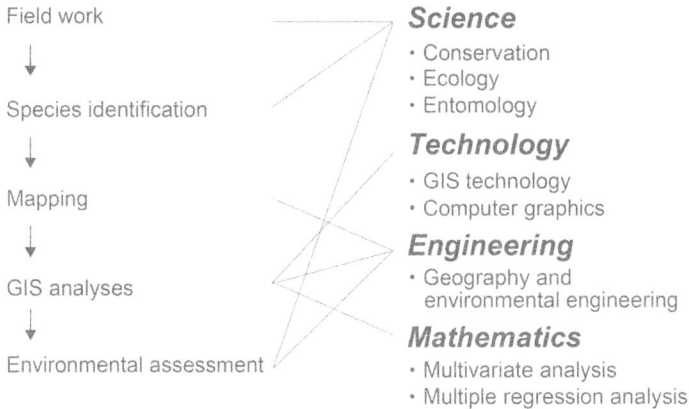

Figure 11.3 The flow of utilization as a teaching aid: environmental education material integrating habitat analyses and environmental assessment using GIS.

The selection of the study site and target organism groups is key to the development of this educational material. It is essential that the study area includes various natural and artificial environments. The target taxonomic groups should not have extremely high migratory and dispersal abilities, and the species that occur in each environment should be limited to some extent. Butterflies are known to be a group that is highly related to the environment. From the viewpoint of educational material, butterflies have the following advantages: (1) they can be easily identified at the naked-eye level; (2) most species are diurnal; (3) various identification support tools, such as illustrated books, are available; and (4) the number of species is appropriate. Butterflies are also a suitable taxon group for studying the relationship between the occurrence of each species and vegetation because many species have a limited range of food plants. In addition to butterflies, dragonflies, which use both aquatic and terrestrial habitats in their life history, are also suitable taxa for understanding the relationship between the occurrence of each species and its environment. The Higashi-hiroshima Campus of Hiroshima University is a suitable field for analyzing the relationship between each butterfly species and its environment as it is home to various environments and the species composition of butterflies in each environment tends to be consistent (Tokuno & Tomikawa, 2022). In this study, the species composition of butterflies on the Higashihiro-shima Campus of Hiroshima University was clarified using the root census method and the butterflies inhabiting the campus as an example. Through ArcGIS Pro, a GIS software provided by ESRI Japan, we created, integrated, and analyzed data on the confirmed points of occurrence obtained by the root census method and the environmental data in the surrounding areas. In addition, we analyzed the habitat of butterflies on the Higashihiro-shima Campus based on the environmental information obtained from satellite image data and attempted to verify the habitat analysis method using satellite images. Furthermore, we examined the possibility of using the obtained data as environmental education material from the viewpoint of environmental assessment.

In this teaching material, students survey the occurrence of butterflies on the university campus through fieldwork, recorded environmental data for each species, and conducted habitat analyses using a GIS analysis software (ArcGIS Pro, ESRI, Japan) to clarify the habitat characteristics of each species (Figure 11.2B). These results can subsequently be used as environmental education material to discuss the environment and species that should be conserved from the perspective of environmental assessment. For example, an activity could discuss which environment and species should be prioritized for conservation, assuming that a new highway or railroad is to be built in the study area.

GIS should be studied in lower secondary school social studies and upper secondary school geography (MEXT, 2017c, 2018c). Studies have also been conducted on the development and practice of interdisciplinary lessons in geography and biology (Kurabayashi et al., 2021). In this teaching material, the species diversity of butterflies was analyzed using GIS, which is taught in social studies. The results were used in the study of ecosystem conservation taught in upper secondary school biology, making it possible to design lessons that link science and social studies. This learning program is expected to be used not only in Japan but also in various countries and regions because the same education can be practiced with different target animals and regions.

Conclusion and implementation

This study provides an overview of STEM/STEAM education related to biology in Japan by reviewing research and educational practices related to STEM/STEAM education focused on biological sciences in Japan. Through an analysis of the curriculum guidelines, we found STEM/STEAM-related statements throughout the curriculum guidelines of biology education in elementary, middle, and upper secondary school. The curricula of the Guidelines for the Course of Study for upper secondary schools also emphasize the "development of qualities and abilities from a cross-curricular perspective," which is consistent with the concept of STEM/STEAM. In Japan, practical research on STEM/STEAM education in the biological sciences has been conducted by adopting various approaches that utilize IMA, simulation, and models and are based on PBL to varying degrees.

It is believed that STEM/STEAM education in biological sciences in Japan is lacking due to the insufficient development of teaching material; therefore, we proposed examples of our recent work in developing STEAM teaching material using biotechnology and GIS. In the development of educational material for "DNA barcoding and molecular phylogenetic analysis of mammals using meat" through biotechnology, students will use familiar ingredients sold in supermarkets to identify species and infer phylogenetic relationships through genetic analysis. This material covers a wide range of topics of biology education in upper secondary school, from the principles and methods of genetic analysis to the evolutionary history of specific mammals. Furthermore, as this learning program targets mammals that are used as meat worldwide, similar education can be implemented in various countries and regions outside of Japan. However, the analysis material is a piece of meat and does not provide information on the morphology, feeding habits, and habitat of individual animal species. Therefore, it is important that taxonomic and ecological information on animal species is provided and learned in advance.

In a study on the development of environmental education material using GIS, environmental education material integrating habitat analysis and environmental assessment using GIS was developed using butterflies that appear on a university campus as an example. This material is excellent because it provides a comprehensive study of field survey methods, methods, and principles of environmental analysis using GIS and specific measures for environmental assessment. Furthermore, this learning program is expected to be used not only in Japan but also in various other countries and regions as the same education can be practiced with different target animals and regions. However, the software used for GIS analysis is not yet widely used in many upper secondary schools, and the number of teachers who can teach the analysis is limited, which limits the number of schools able to implement it. Because GIS has recently become a required part of social studies in middle school and geography in upper secondary school, designing lessons in conjunction with science and social studies will help to overcome technical problems while achieving cross-curricular learning.

Biological science is one of the fields most closely related to human life, and remarkable progress in biology research in recent years has greatly improved the quality and quantity of teaching content in biology education. With the development of science and technology, textbooks have started to include many results of cutting-edge research, especially in genetics. In the future, an understanding of biology, the science and technology that led to these results, and computational methods such as statistical processing will be essential to understand modern biology. In this respect, STEM/STEAM perspectives and cross-curricular learning will become increasingly important.

Acknowledgment

This study was partially supported by the Japan Society for the Promotion of Science KAKENHI Grant Numbers JP21H00919 (JP23K20744), JP22K06373.

References

Arai, K. (2018). Past and future of STEM education. *STEM Education*, *1*, 3–7.
Arai, K. (2019). STEM education and new guidelines for teaching. *Research Institute for Mathematics and Science Education*, *26*, 12–16.

Ari, A. G., & Meço, G. (2021). A new application in biology education: development and implementation of Arduino-supported STEM activities. *Biology, 10*, 506.

Bradley-Levine, J., & Mosier, G. (2014). *Literature review on project-based learning*. University of Indianapolis Center of Excellence in Leadership of Learning. http://1stmakerspace.com.s3.amazonaws.com/Resources/PBL-Lit-Review_Jan14.2014.pdf

Hasanah, U. (2020). Key definitions of STEM education: Literature review. *Interdisciplinary Journal of Environmental and Science Education, 16*(3), e2217.

Iida, Y., & Katayama, N. (2021). Measuring the rate of photosynthesis of seaweeds and other organisms using Easysense. *The Heredity, 75*(5), 456–463.

Isaji, K., & Matsumoto, S. (2005). Development of teaching material for DNA extraction method practicable at high school. *Annual report of the Faculty of Education, Gifu University. Educational Research, 7*, 69–78.

Isozaki, T., & Isozaki, T. (2021). Theoretical research for establishing a Japanese-style STEM education: Analysis from a comparative historical point of view. *Journal of Science Education in Japan, 45*(2), 142–154. https://doi.org/10.14935/jssej.45.142

Jones, T. C., & Laughlin, T. F. (2010). PopGen fishbowl: A free online simulation model of microevolutionary processes. *The American Biology Teacher, 72*(2), 100–103. https://doi.org/10.1525/abt.2010.72.2.9

Kazama, T., Yamanoi, T., & Takemura, M. (2014). Development of vegetable-based methods for learning knowledge related to the molecular phylogenetic tree in a Japanese upper secondary school biology course. *Journal of Research in Science Education, 54*(3), 319–334.

Kitagaki, K., & Andoh, H. (2007). Mat lush as a teaching material II. *JSSE Research Report, 23*(2), 13–16.

Kosaka, N., & Kumano, Y. (2021). A study of changes in US high school biology textbooks: Focusing on the impact of the STEM (Science, Technology, Engineering and Mathematics) education reform by the Next Generation Science Standards (NGSS). *Japanese Journal of Biological Education, 62*(3), 128–139.

Kurabayashi, M., Takahashi, E., Fukaya, S., & Takemura, M. (2021). Development and practice of inter-curricular lessons on geography and biology using GIS, through the creation and analysis of cherry blossom maps. *The New Geography, 69*(2), 54–68.

Matsubara, K., & Kosaka, M. (2017). A discussion of STEM education and questions fostering competencies in the Japanese curriculum. *Journal of Science Education in Japan, 41*(2), 150–160. https://doi.org/10.14935/jssej.41.150

Ministry of Education, Culture, Sports, Science, and Technology (MEXT). (2017a). *Sho Gakko Gakushu Shido Yoryo Kaisetsu Rika Hen [Commentary on the courses of study for elementary schools, science]*. https://www.mext.go.jp/content/20211020-mxt_kyoiku02-100002607_05.pdf

Ministry of Education, Culture, Sports, Science, and Technology (MEXT). (2017b). *Chu Gakko Gakushu Shido Yoryo Kaisetsu Rika Hen [Commentary on the courses of study for lower secondary schools, science]*. https://www.mext.go.jp/content/20210830-mxt_kyoiku01-100002608_05.pdf

Ministry of Education, Culture, Sports, Science, and Technology (MEXT). (2017c). *Chu Gakko Gakushu Shido Yoryo Kaisetsu Shakai Hen [Commentary on the courses of study for lower secondary schools, social studies]*. https://www.mext.go.jp/component/a_menu/education/micro_detail/__icsFiles/afieldfile/2019/03/18/1387018_003.pdf

Ministry of Education, Culture, Sports, Science, and Technology (MEXT). (2017d). *Chu Gakko Gakushu Shido Yoryo Kaisetsu Hokentaiiku Hen [Commentary on the courses of study for lower secondary schools, health and physical education].* https://www.mext.go.jp/content/20210113-mxt_kyoiku01-100002608_1.pdf

Ministry of Education, Culture, Sports, Science, and Technology (MEXT). (2018a). *Koto Gakko Gakushu Shido Yoryo [Upper secondary school curriculum guidelines].* https://erid.nier.go.jp/files/COFS/h30h/index.htm

Ministry of Education, Culture, Sports, Science, and Technology (MEXT). (2018b). *Koto Gakko Gakushu Shido Yoryo Kaisetsu Rika Hen Risuu Hen [Commentary on the courses of study for upper secondary schools, sciences, science and mathematics].* https://www.mext.go.jp/content/20211102-mxt_kyoiku02-100002620_06.pdf

Ministry of Education, Culture, Sports, Science, and Technology (MEXT). (2018c). *Koto Gakko Gakushu Shido Yoryo Kaisetsu Chiri Rekishi Hen [Commentary on the courses of study for upper secondary schools, geography, history].* https://www.mext.go.jp/content/20220802-mxt_kyoiku02-100002620_03.pdf

Ministry of Education, Culture, Sports, Science, and Technology (MEXT). (2018d). *Koto Gakko Gakushu Shido Yoryo Kaisetsu Komin Hen [Commentary on the courses of study for upper secondary schools, civics].* https://www.mext.go.jp/content/20211102-mxt_kyoiku02-100002620_04.pdf

Ministry of Education, Culture, Sports, Science, and Technology (MEXT). (2018e). *Koto Gakko Gakushu Shido Yoryo Kaisetsu Hoken Taiiku Hen, Taiiku Hen [Commentary on the courses of study for upper secondary schools, health and physical education, physical education].* https://www.mext.go.jp/content/1407073_07_1_2.pdf

Ministry of Education, Culture, Sports, Science, and Technology (MEXT). (2018f). *Koto Gakko Gakushu Shido Yoryo Kaisetsu Katei Hen [Commentary on the courses of study for upper secondary schools, home economics].* https://www.mext.go.jp/content/1407073_10_1_2.pdf

Munn, M., Skinner, P. O., Conn, L., Horsma, H. G., & Gregory, P. (1999). The involvement of genome researchers in high school science education. *Genomic Research, 9*(7), 597–607.

National Science Foundation. (2013). Graduate Research Fellowship Program (GRFP). NSF 14-590.

Okumura, J., & Kumano, Y. (2016). Action research on expansion of biological knowledge and changes in the scientific and engineering thought processes in high school students in the learning of the avian embryonic experiment based on a Bio-STEM perspective. *Journal of Science Education in Japan, 40*(1), 21–29. https://doi.org/10.14935/jssej.40.21

Okumura, J., & Kumano, Y. (2018) The action research on the changes of scientific thinking with Bio-STEM education at high school biology. *Bulletin of the Center for Educational Research and Teacher Development Shizuoka University*, 28, 125–133. http://doi.org/10.14945/00024668

Sato, A., Yamanoi, T., Kashiwagi, J., & Aoki, Y. (2018). Android tablet applications for lower secondary education to promote understanding of Mendelian inheritance and dominant and recessive traits. *Japanese Journal of Biological Education, 59*(2), 64–74. https://doi.org/10.24718/jjbe.59.2_64

Sawyer, R. K. (2014). Introduction: The new science of learning. In R. K. Sawyer (Ed.), *The Cambridge handbook of the learning science* (2nd ed., pp. 1–18). Cambridge University Press.

Sengoku, K., Tabata, T., Saito, M., & Monma, M. (1993). Sequencing of mitochondrial cytochrome b genes for the identification of meat species. *Journal of Animal Science and Technology, 65*(6), 571–579.

Shingai, Y., Aoki, R., Kobayashi, W., Takeda, Y., Kusunoki, F., Mizoguchi, H., Sugimoto, M., Funaoi, H., Yamaguchi, E., & Inagaki, S. (2021). *Satoyama* management game, a learning support system for acquiring problem solving skills in STEM education: An examination of the learning effect when dealing with multiple *Satoyama*. *Journal of Science Education in Japan*, *45*(2), 112–127. https://doi.org/10.14935/jssej.45.112

Sonoyama, H. (2020). Development of teaching material that can be completed by students ding a single class while effectively fostering their comprehensive understanding of the function of restriction enzymes. *Journal of Research in Science Education*, *61*(1), 167–173. https://doi.org/10.11639/sjst.19074

Sousa, D. A., & Pilecki, T. (2013). *From STEM to STEAM: Using brain-compatible strategies to integrate the arts.* Corwin Press.

Takagi, S., Tokoro, Y., Fujiwara, Y., & Yamashita, S. (2001). Development of home economics teaching materials to enhance students' understanding of the role of enzymes in Living environment: Kinetic analysis of Protease activity by viscometry and its activity relating to food in life. *Journal of Science Education in Japan*, *25*(1), 35–43. https://doi.org/10.14935/jssej.25.35

Tamura, K., Stecher, G., & Kumar, S. (2021) MEGA11: Molecular Evolutionary Genetics Analysis Version 11. *Molecular Biology and Evolution*, *38*(7), 3022–3027. https://doi.org/10.1093/molbev/msab120

Taniguchi, D. (2023). "Classroom practice on the properties of dilute solutions, challenges in teaching using a self-made osmotic pressure experimental apparatus," Keirinkan. Retrieved February 7, 2023, from https://www.shinko-keirin.co.jp/keirinkan/kou/science/kagaku-jissen/202101/

Tokuno K, & Tomikawa, K. (2022). Butterfly fauna and its habitat preference in Higashihiroshima Campus of Hiroshima University. *Bulletin of the Hiroshima University Museum*, *14*, 67–74.

Tomikawa, K., & Mukuda, T. (2023). Understanding osmotic regulation of body fluids using *Neocaridina denticulate*. *The Heredity*, *77*(2), 64–68.

Tsujiai, H., & Hasegawa, H. (2020). A study of the concept of "A" in STEAM education. *Journal of Science Education in Japan*, *44*(2), 93–103. https://doi.org/10.14935/jssej.44.93

Ueno, Y., & Kurihara, M. (2014). An attempt to evaluate and map the habitat of organisms on a broad scale using GIS and habitat suitability models. *Civil Engineering Journal*, *56*, 22–25.

Yamanoi, T., Takemura, M., Sakura, O., & Kazama, T. (2012). Development and evaluation of an activity to teach molecular phylogeny, deep time and classification systems for Japanese high school students. *Asian Journal of Biology Education*, *6*, 13–25. https://doi.org/10.57443/ajbe.6.0_13

Yamashita, S., & Nomura, K. (2017). Development of STEM science lessons using artificial heart valve. *Japan Society of Science Education, Annual Convention Memoirs*, *41*, 85–88.

Index

Note: Locators in *italics* represent figures and **bold** indicate tables in the text.

For Product Safety Concerns and Information please contact our EU
representative GPSR@taylorandfrancis.com
Taylor & Francis Verlag GmbH, Kaufingerstraße 24, 80331 München, Germany

www.ingramcontent.com/pod-product-compliance
Lightning Source LLC
Chambersburg PA
CBHW060307220326
41598CB00027B/4264